TABLE OF CONTENTS

Dominoes (Sums 1-4)	1
Coloring Picture (Sums 1-4)	2
Maze (Sums 1-6)	3
Number Pairs (Sums 3,4)	4
Matching Equal Equations (Sums 1-8)	5
Addition Square Puzzles (Sums 1-4)	6
Coloring Picture (Sums 5-8)	7
Writing Equations (Sums 5-8)	8
Number Pairs (Sums 5,6)	9
Number Pairs (Sums 7,8)	10
Maze (Sums 5-8)	11
Magic Squares (Sums 2-9)	12
Addition Maze (Sums 5-8)	13
Addition Square Puzzles (Sums 5-8)	14
Coloring Picture (Sums 9-12)	15
Writing Equations (Sums 9-12)	16
Number Pairs (Sums 9,10)	17
Number Pairs (Sums 11,12)	18
Connecting Addends (Sums 9-12)	19
Maze (Sums 9-12)	20
Magic Squares (Sums 1-9)	21
Addition Maze (Sums 9-12)	22
Sumflowers (Sums 3-14)	23
Hidden Stars (Sums 4-18)	24
Satellite Sums (Sums 7,9,10)	25
Addition Square Puzzles (Sums 9-12)	26
Sum Apples (Sums 9-12)	27
Coloring Picture (Sums 13-16)	28
Number Pairs (Sums 13,14)	29
Number Pairs (Sums 15,16)	30
Connecting Addends (Sums 13-16)	31
Maze (Sums 13-16)	32
Magic Squares (Sums 1-12)	33
Sumflowers (Sums 4-23)	34
Matching Equal Equations (Sums 12-15)	35
Satellite Sums (Sums 8,11,14,16)	36
Super Satellite Sums (Sums 11,12,13,15)	37
Addition Square Puzzles (Sums 13-16)	38
Pyramid Sums (Sums 9-13)	39
Chart Sums (Sums 3-10)	40
Coloring Picture (Sums 17-20)	41
Connecting Addends (Sums 17-20)	42
Maze (Sums 17-21)	43
Magic Squares (Sums 2-19)	44
Magic Squares (Sums 1-9)	45
More Chart Sums (Sums 7-15)	46
Make Your Own Chart	47
Sumflowers (Sums 12-25)	48
Dot To Dot (Sums 0-24)	49
Dot To Dot (Sums 0-26)	50
Hidden Stars (Sums 2-22)	51
Super Satellite Sums (Sums 14,17,20,24)	52
XYZ Paths (Sums 18-20)	53
Pyramid Sums (Sums 23-25)	54
Two Letter Words (Sums 7-77)	55
Coloring Design (Sums 20-23)	56
Circle Addition (Sums 17-20)	57
Just For Fun (Sums 5-15)	58
Addition Test	59,60
Awards	61,62

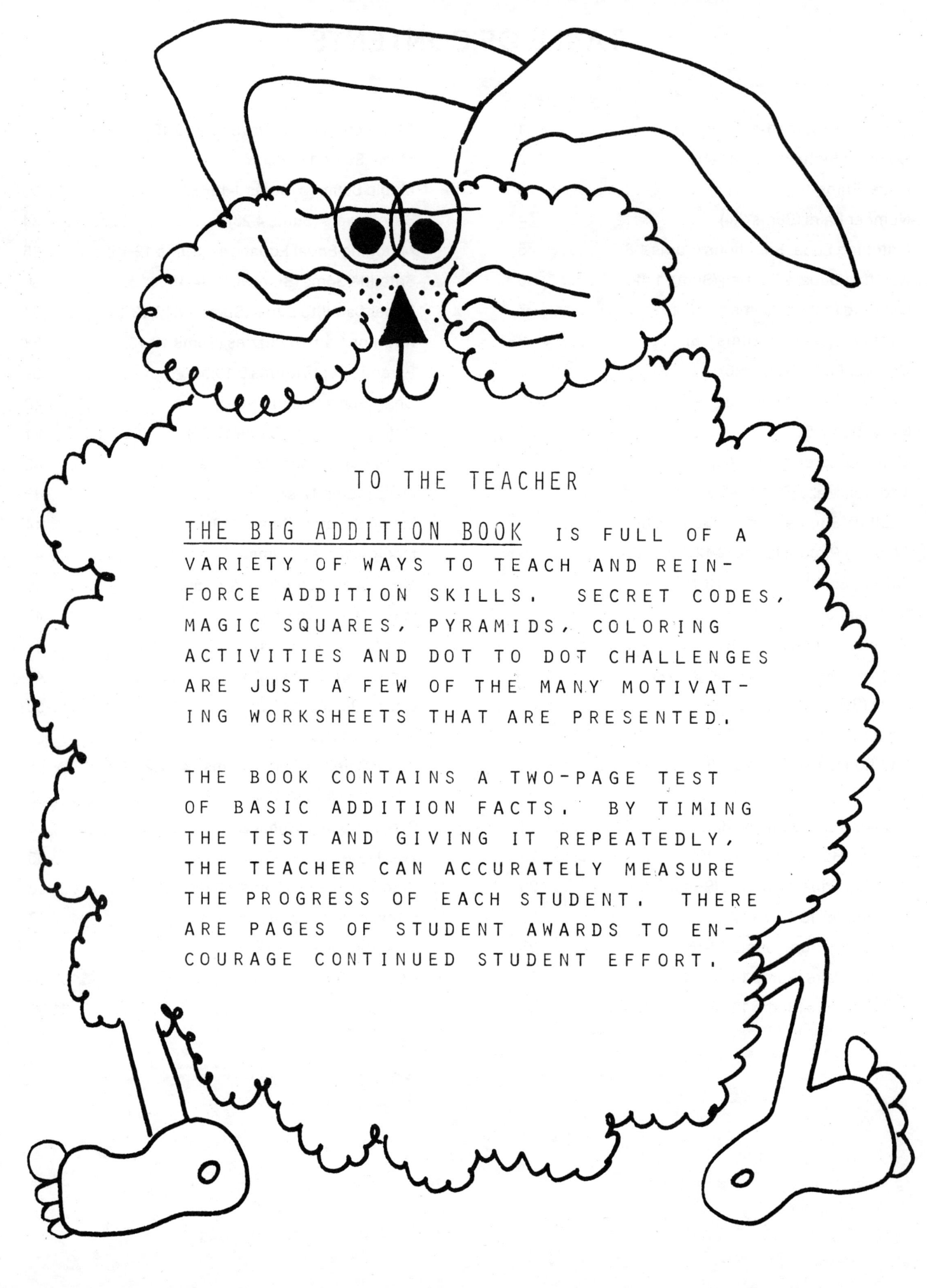

TO THE TEACHER

THE BIG ADDITION BOOK IS FULL OF A VARIETY OF WAYS TO TEACH AND REINFORCE ADDITION SKILLS. SECRET CODES, MAGIC SQUARES, PYRAMIDS, COLORING ACTIVITIES AND DOT TO DOT CHALLENGES ARE JUST A FEW OF THE MANY MOTIVATING WORKSHEETS THAT ARE PRESENTED.

THE BOOK CONTAINS A TWO-PAGE TEST OF BASIC ADDITION FACTS. BY TIMING THE TEST AND GIVING IT REPEATEDLY, THE TEACHER CAN ACCURATELY MEASURE THE PROGRESS OF EACH STUDENT. THERE ARE PAGES OF STUDENT AWARDS TO ENCOURAGE CONTINUED STUDENT EFFORT.

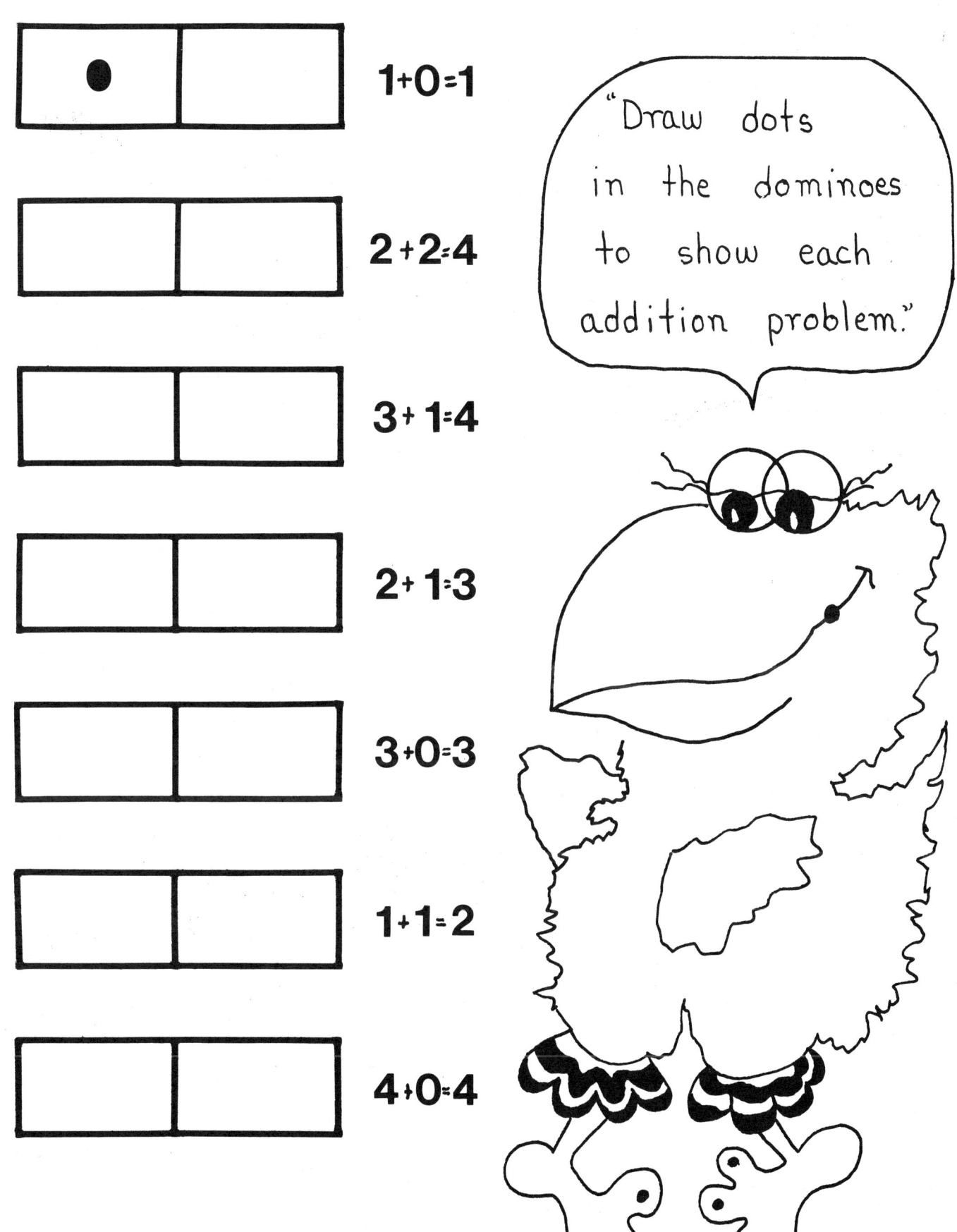

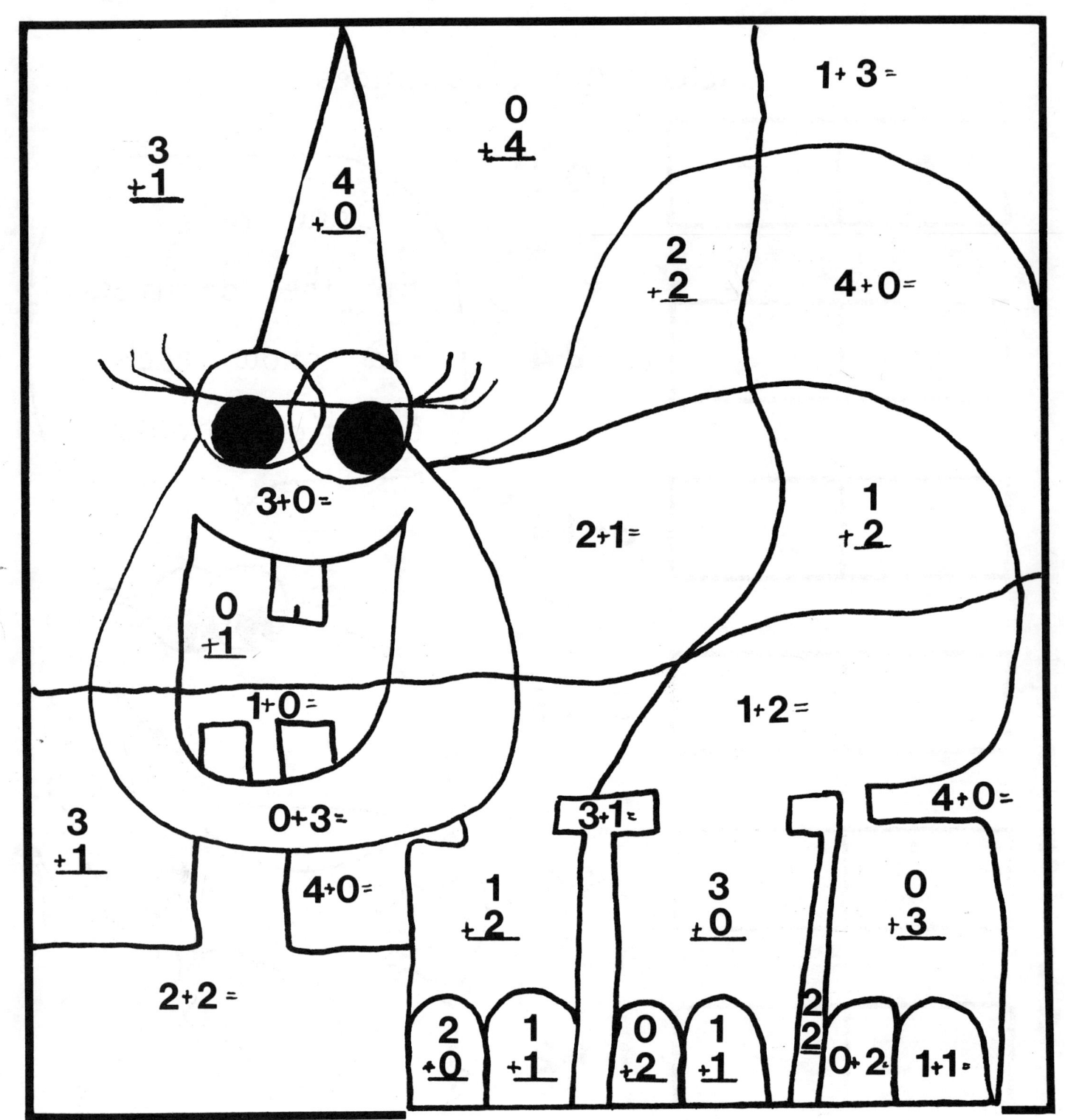

COLORING FUN

COLOR THE SUMS THAT EQUAL (1), PINK.

COLOR THE SUMS THAT EQUAL (2), BROWN.

COLOR THE SUMS THAT EQUAL (3), YELLOW.

COLOR THE SUMS THAT EQUAL (4), PURPLE.

GO! GO! GO! GO!

START AT "GO" AND FINISH AT "STOP." AS YOU PASS THROUGH A CIRCLE, ADD THE NUMBER TO YOUR GROWING TOTAL.

CAN YOU FIND A PATH THAT TOTALS 1?
CAN YOU FIND A PATH THAT TOTALS 3?
CAN YOU FIND A PATH THAT TOTALS 4?
CAN YOU FIND A PATH THAT TOTALS 5?
CAN YOU FIND A PATH THAT TOTALS 6?

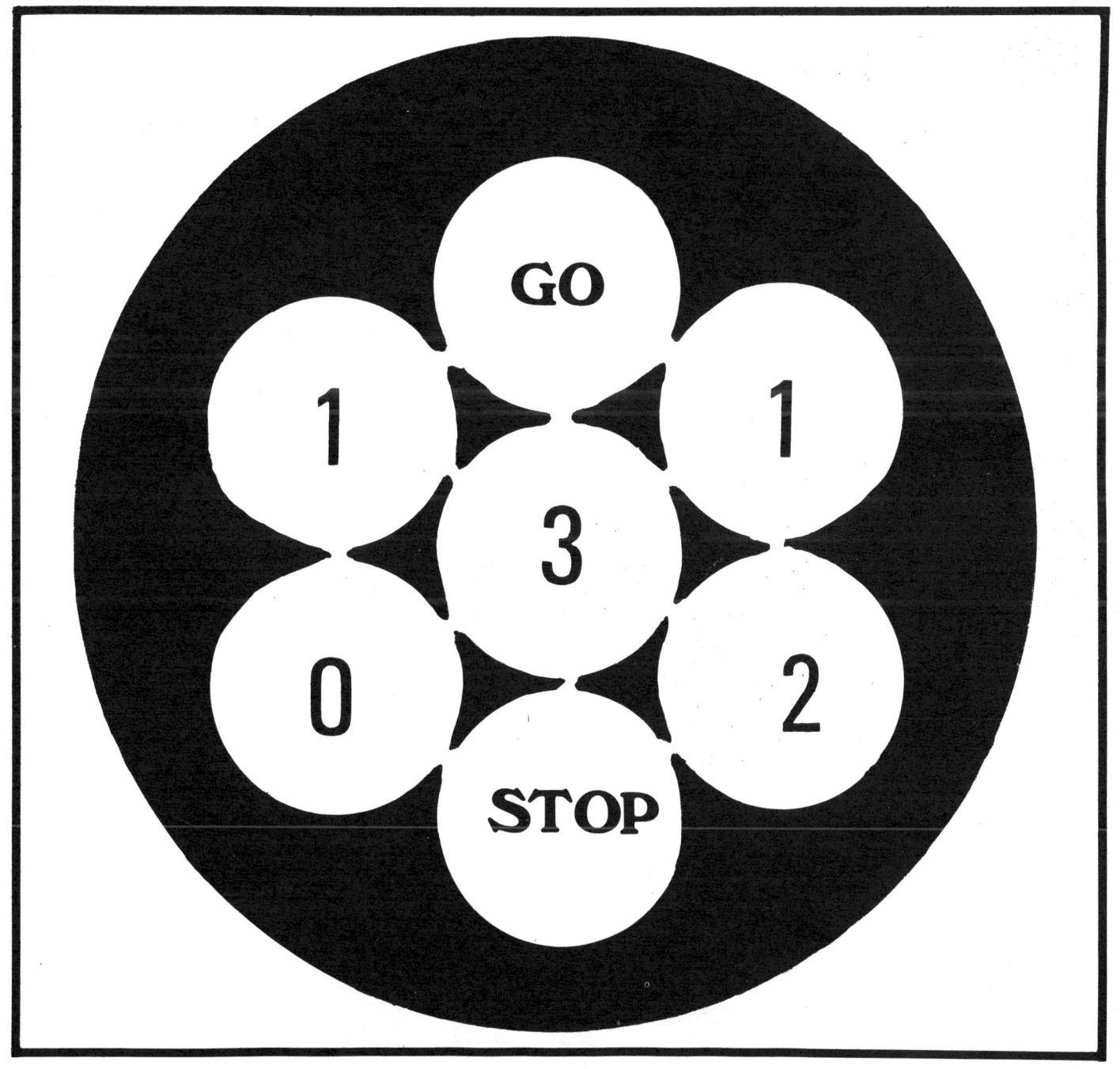

NUMBER PAIRS

0	4	2	4	1	3
0	1	1	3	1	1
1	1	4	2	1	3
4	0	0	3	1	0
1	1	0	4	4	1
0	4	3	1	1	4

PUT A RED CIRCLE AROUND THE NUMBER PAIRS THAT TOTAL THREE (3).

PUT A GREEN CIRCLE AROUND THE NUMBER PAIRS THAT TOTAL FOUR.

DRAW A LINE TO CONNECT EQUATIONS THAT HAVE THE SAME SUMS. THE FIRST ONE IS DONE FOR YOU.

"How many ways can you write 4?"

0 + 4 = 4 _____
_____ _____
_____
_____

2 + 0	3 + 3
4 + 2	1 + 1
0 + 4	3 + 1
4 + 1	3 + 5
3 + 4	3 + 2
4 + 4	6 + 1

0 + 1	2 + 0
1 + 1	1 + 0
0 + 3	3 + 3
2 + 2	2 + 1
5 + 1	4 + 0
0 + 7	3 + 4

CAN YOU DRAW TWO STRAIGHT LINES AND DIVIDE THIS SQUARE SO THAT EACH AREA TOTALS ONE?

CAN YOU DRAW TWO STRAIGHT LINES AND DIVIDE THIS SQUARE SO THAT EACH AREA TOTALS TWO?

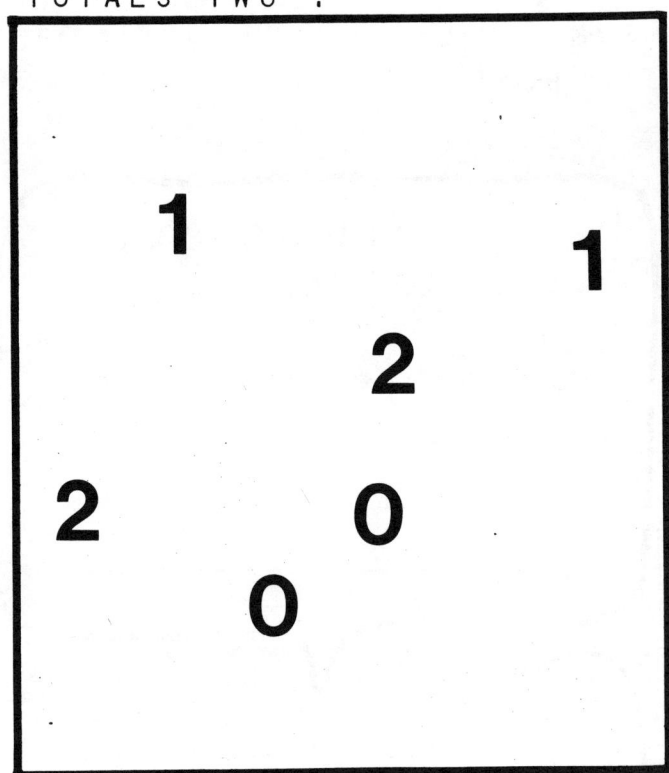

CAN YOU DRAW TWO STRAIGHT LINES AND DIVIDE THIS SQUARE SO THAT EACH AREA TOTALS THREE?

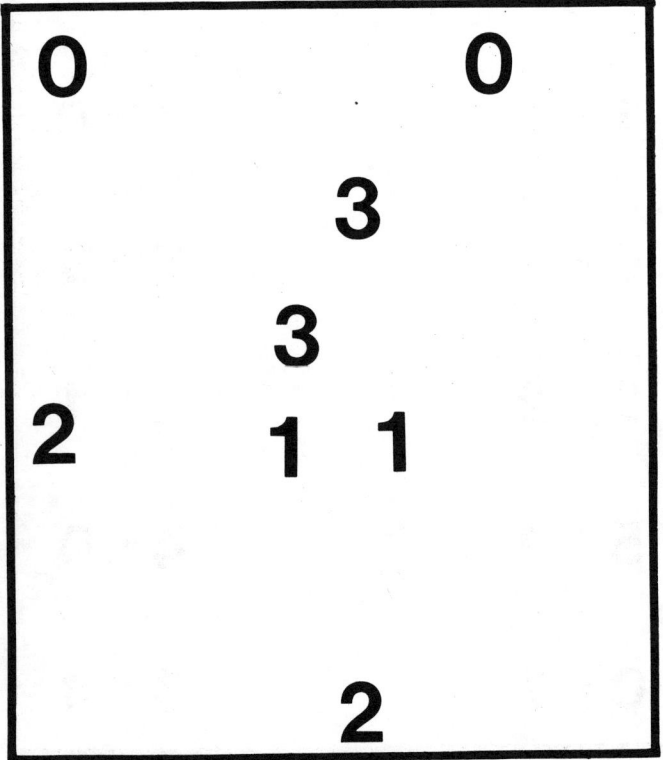

CAN YOU DRAW TWO STRAIGHT LINES AND DIVIDE THIS SQUARE SO THAT EACH AREA TOTALS FOUR?

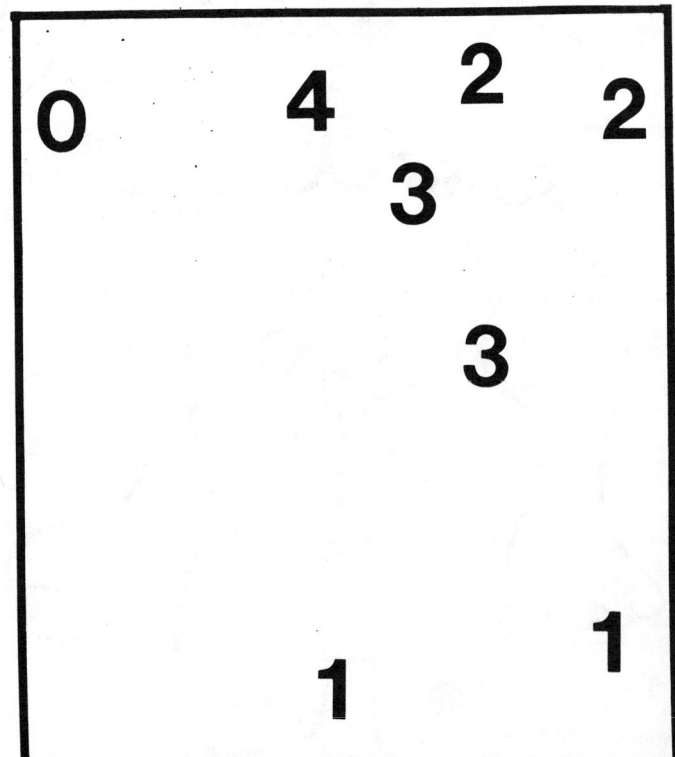

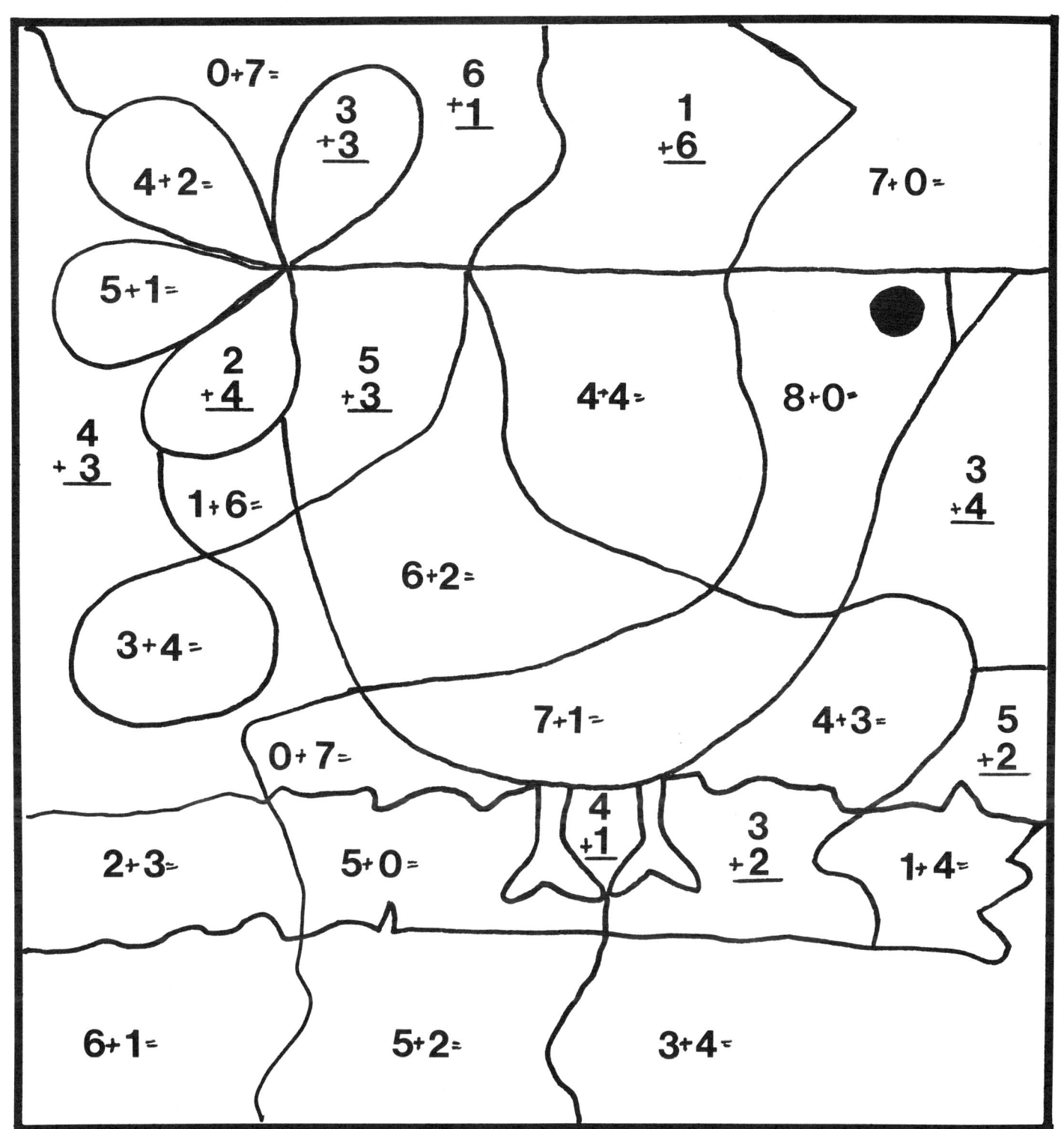

COLORING FUN

COLOR THE SUMS THAT EQUAL FIVE (5), GREEN.
COLOR THE SUMS THAT EQUAL SIX (6), RED.
COLOR THE SUMS THAT EQUAL SEVEN (7), BLUE.
COLOR THE SUMS THAT EQUAL EIGHT (8), YELLOW.
I HAVE COLORED A PICTURE OF A _____.

"Read the big number in each box. Write as many addition equations as you can to equal each number."

5
5+0

6

7

8

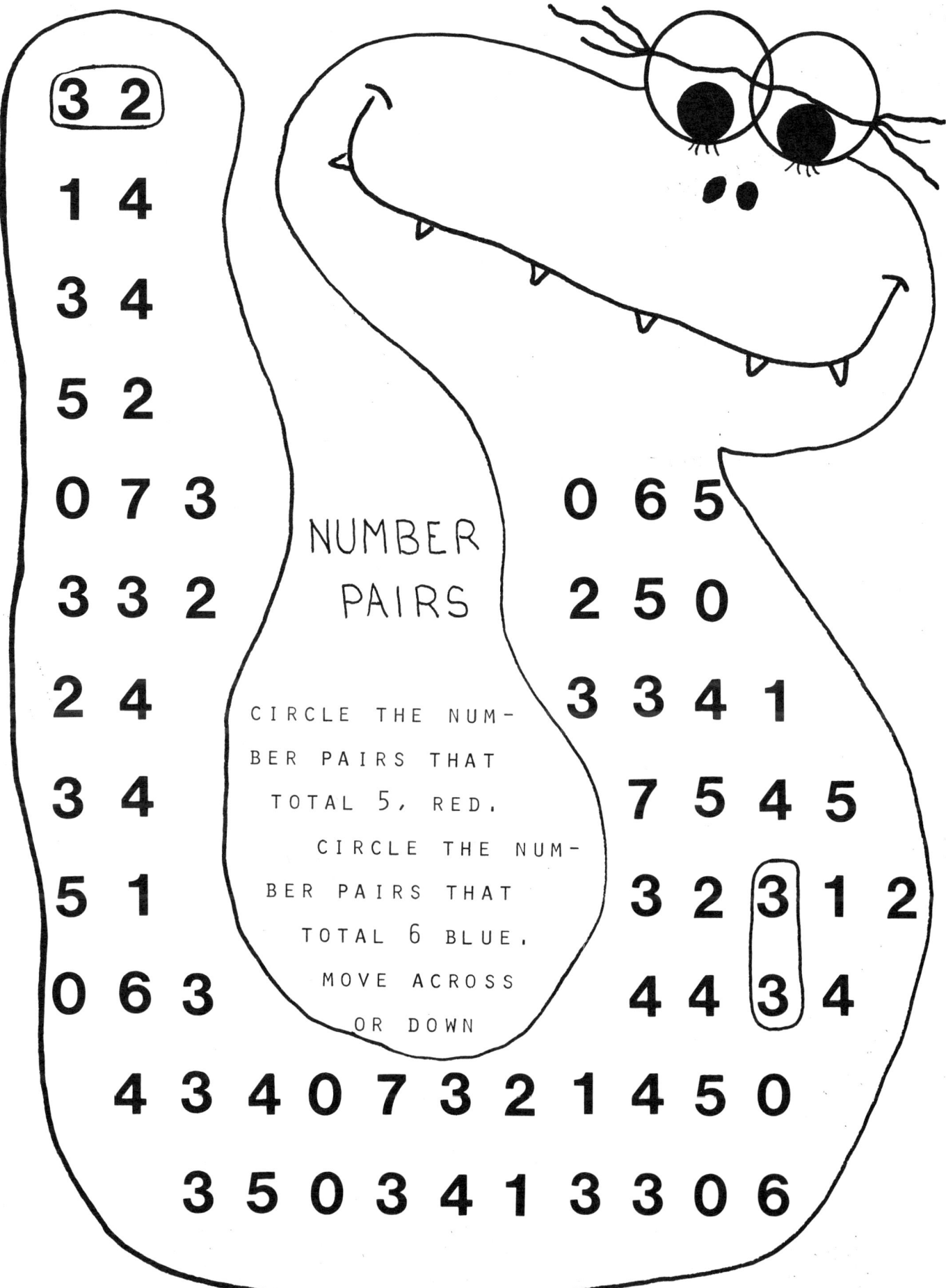

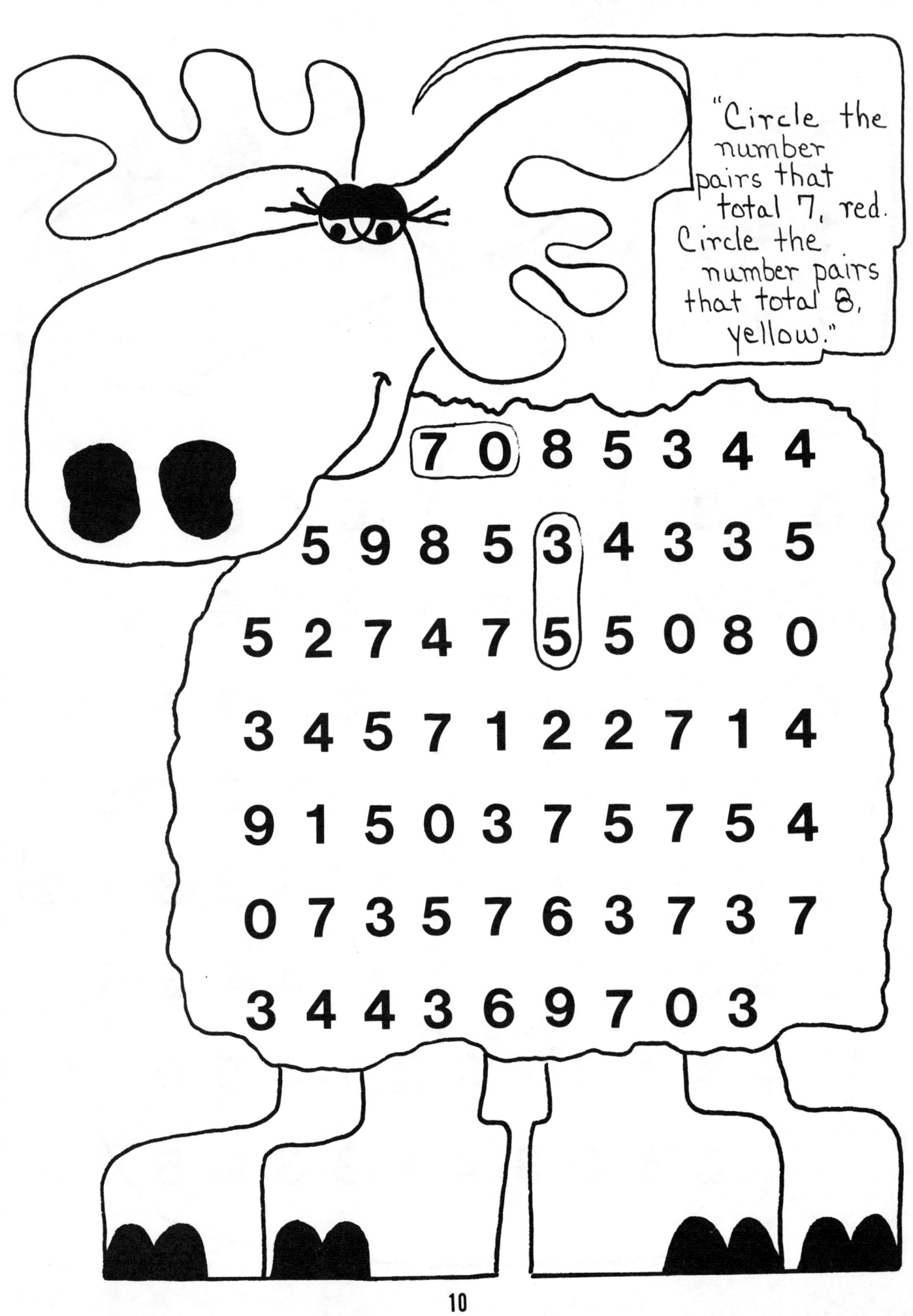

START AT "GO" AND FINISH AT "STOP." AS YOU PASS THROUGH A CIRCLE, ADD THE NUMBER TO YOUR TOTAL.

CAN YOU FIND A TRAIL THAT TOTALS 5 ?
CAN YOU FIND A TRAIL THAT TOTALS 6 ?
CAN YOU FIND A TRAIL THAT TOTALS 7 ?
CAN YOU FIND A TRAIL THAT TOTALS 8 ?

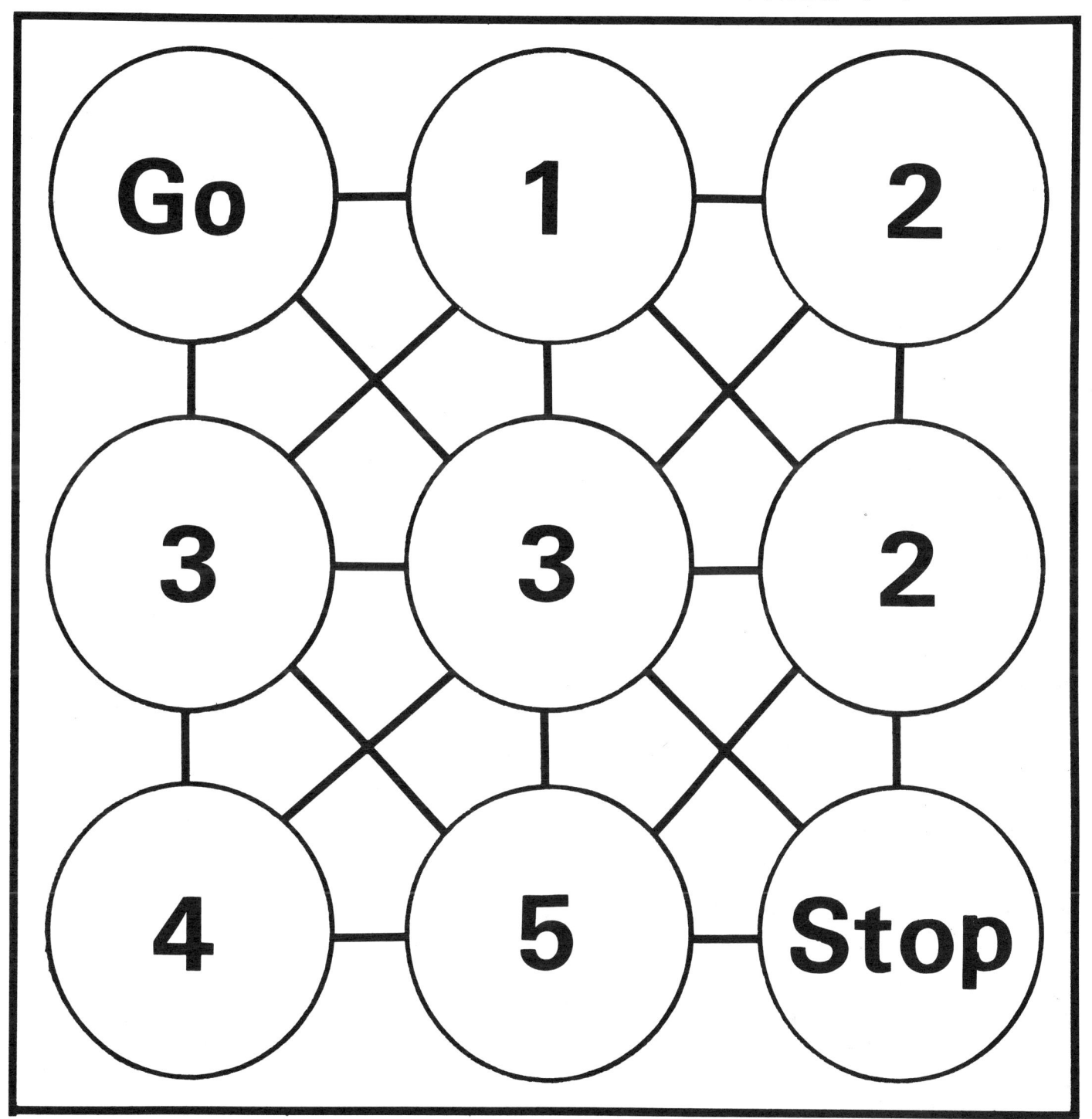

MAGIC SQUARES

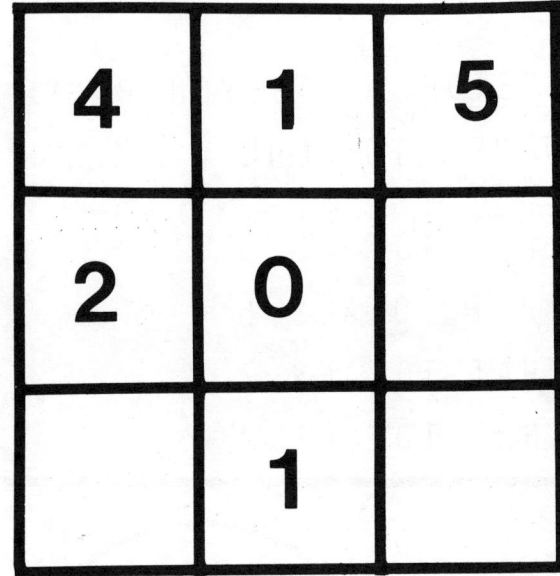

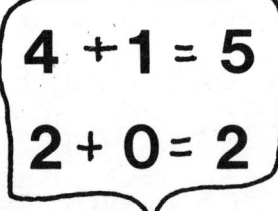

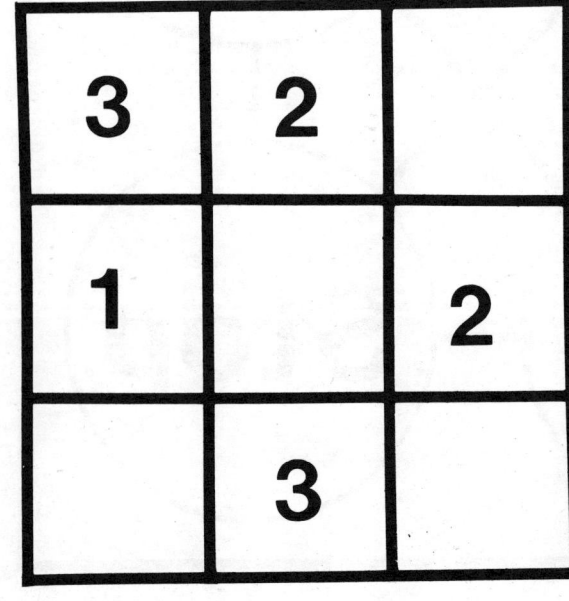

DIRECTIONS: THE FIRST TWO NUMBERS IN EACH ROW EQUAL THE LAST NUMBER. MOVE ACROSS AND DOWN.

CAN YOU DRAW 2 STRAIGHT LINES THAT WILL DIVIDE THE SQUARE, SO THAT EACH AREA TOTALS 5 ?

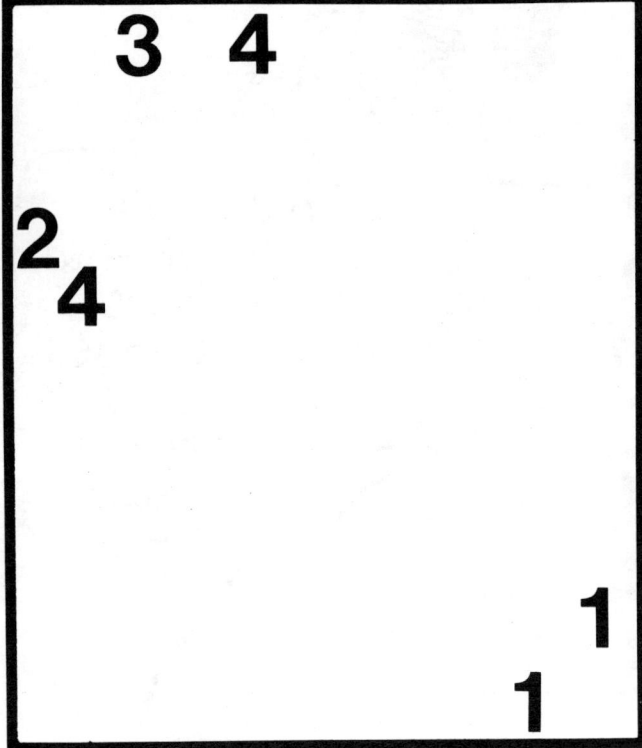

CAN YOU DRAW 2 STRAIGHT LINES THAT WILL DIVIDE THE SQUARE, SO THAT EACH AREA TOTALS 6 ?

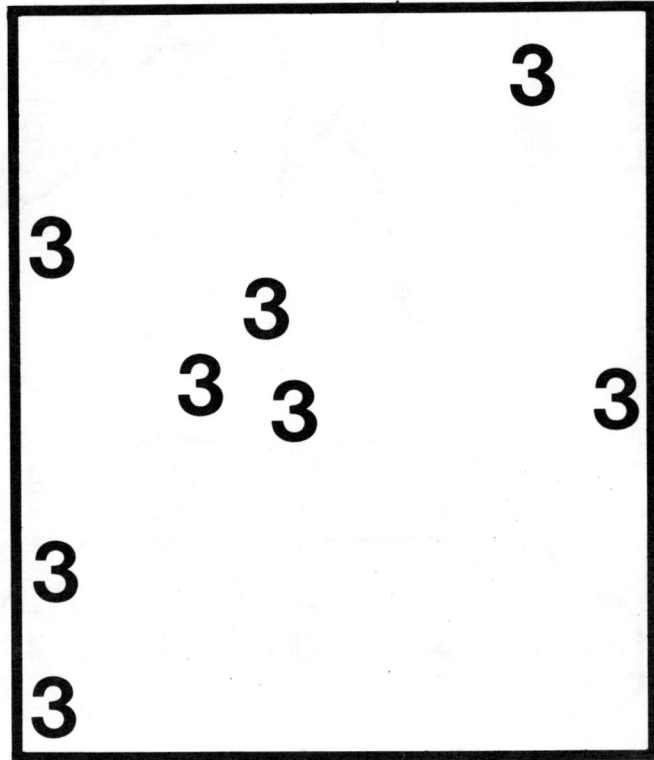

CAN YOU DRAW 2 STRAIGHT LINES THAT WILL DIVIDE THE SQUARE, SO THAT EACH AREA TOTALS 7 ?

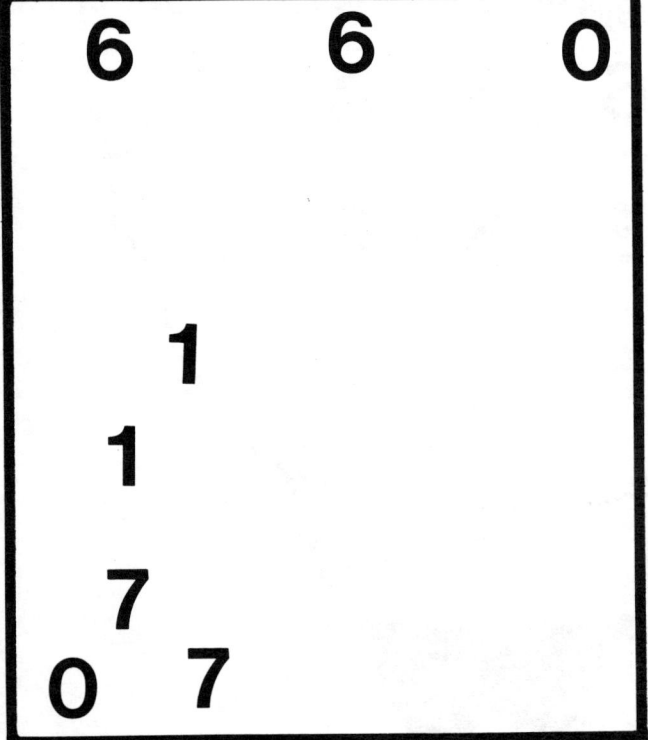

CAN YOU DRAW 2 STRAIGHT LINES THAT WILL DIVIDE THE SQUARE, SO THAT EACH AREA TOTALS 8 ?

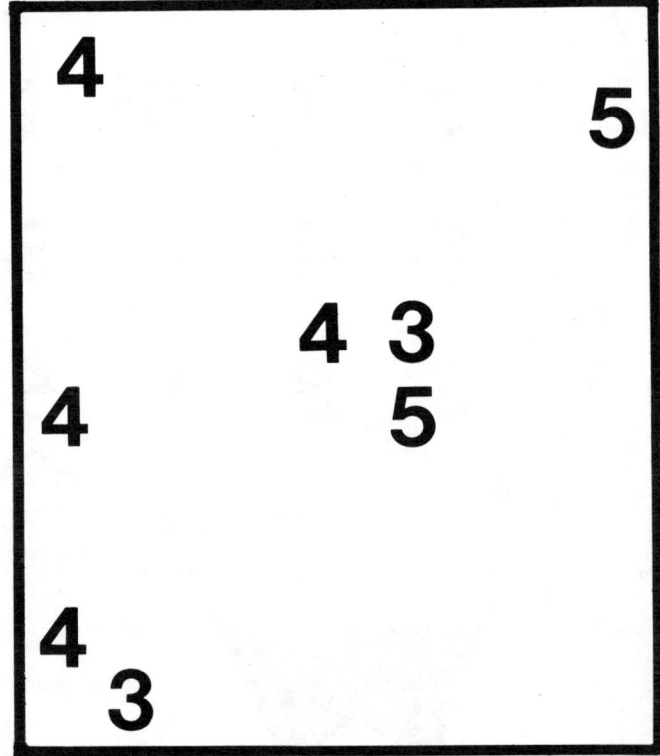

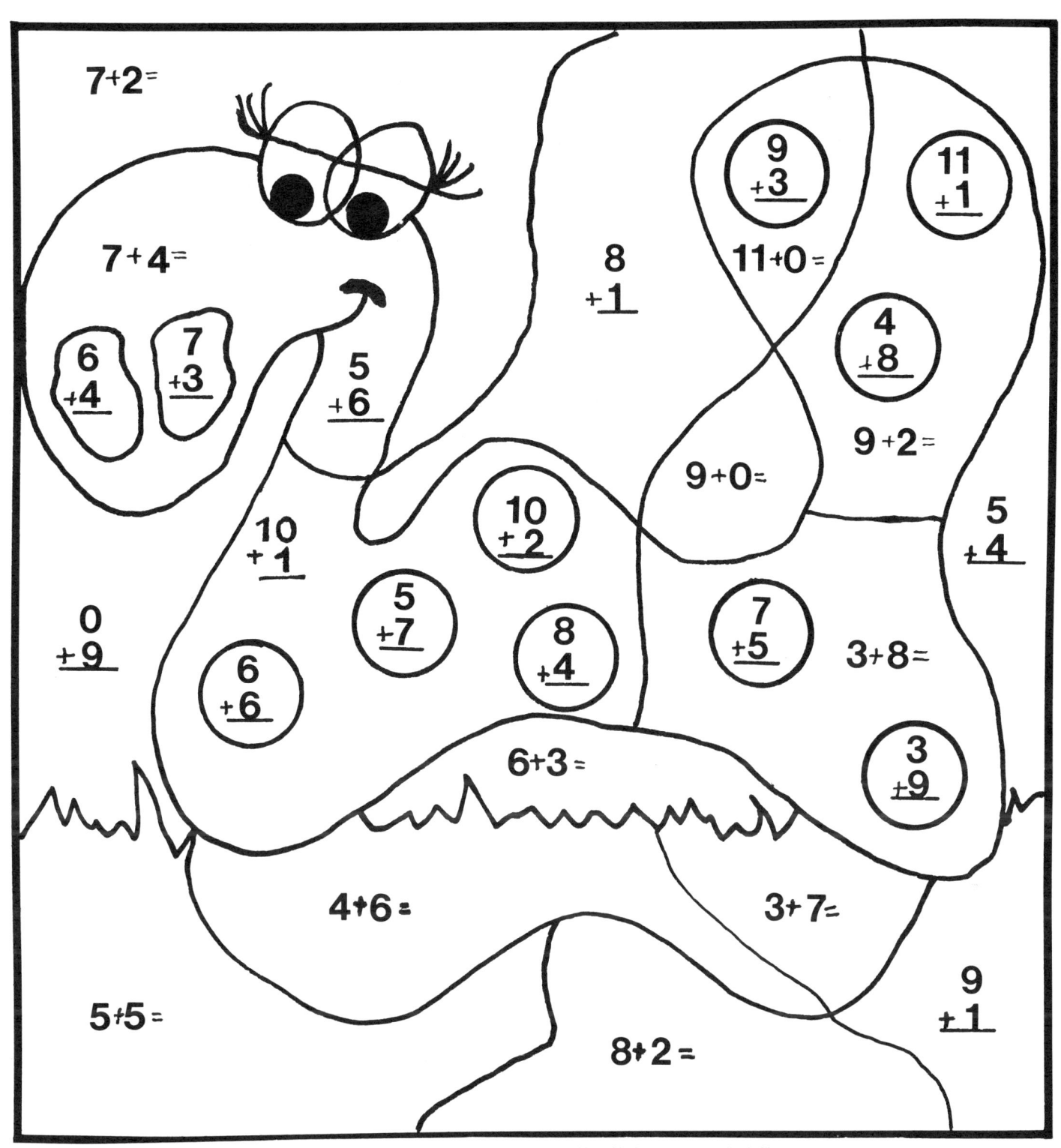

COLOR THE SUMS THAT EQUAL NINE (9), PINK.

COLOR THE SUMS THAT EQUAL TEN (10), GREEN.

COLOR THE SUMS THAT EQUAL ELEVEN (11), ORANGE.

COLOR THE SUMS THAT EQUAL TWELVE (12), BLACK.

I HAVE COLORED A PICTURE OF A _____.

9	**10**
9 + 0 = 9	

11	**12**

"Look at the big number in each box. Write as many equations as you can to equal that number."

WRITE THE REVERSE OF EACH OF THE EQUATIONS BELOW. THE FIRST ONE IS DONE FOR YOU.

1 + 7 = 7 + 1

6 + 3

8 + 2

5 + 7

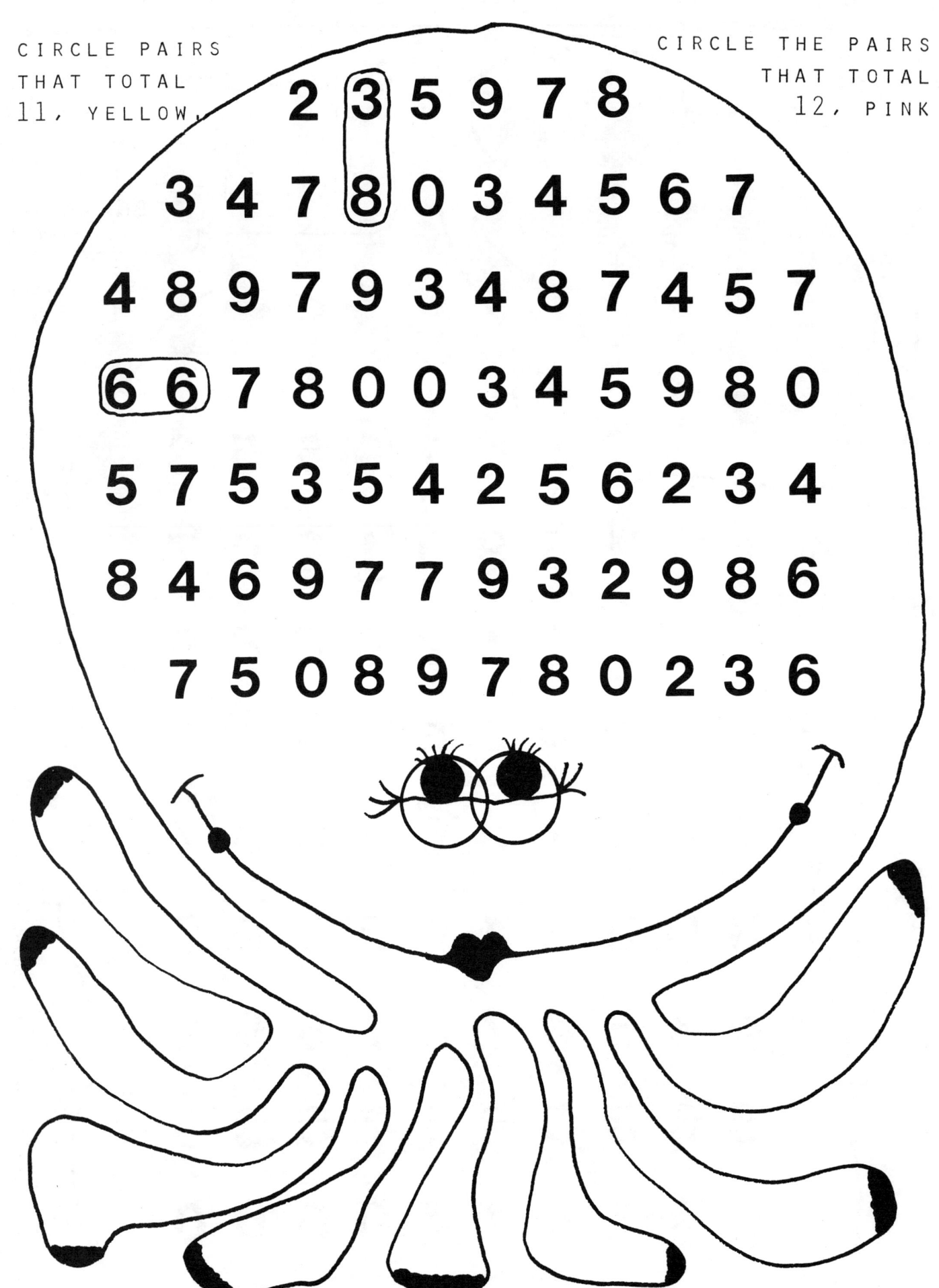

DRAW A LINE TO CONNECT EACH PAIR OF NUMBERS THAT EQUALS 9.		DRAW A LINE TO CONNECT EACH PAIR OF NUMBERS THAT EQUALS 10.	
3	9	5	7
7	8	6	6
2	1	4	5
8	2	3	4
6	0	7	1
1	6	8	2
9	3	2	8
0	7	9	3

(A dashed line connects the 7 on the left with the 2 on the right in the first box.)

DRAW A LINE TO CONNECT EACH PAIR OF NUMBERS THAT EQUALS 11.		DRAW A LINE TO CONNECT EACH PAIR OF NUMBERS THAT EQUALS 12.	
7	6	6	3
4	7	4	9
5	4	8	7
6	8	9	4
3	5	3	5
8	9	5	6
9	3	7	2
2	2	10	8

CAN YOU FIND A TRAIL THAT TOTALS 9 ?
CAN YOU FIND A TRAIL THAT TOTALS 10 ?
CAN YOU FIND A TRAIL THAT TOTALS 11 ?
CAN YOU FIND A TRAIL THAT TOTALS 12 ?
START AT "GO", END AT "STOP." ADD AS YOU PASS NUMBERS.

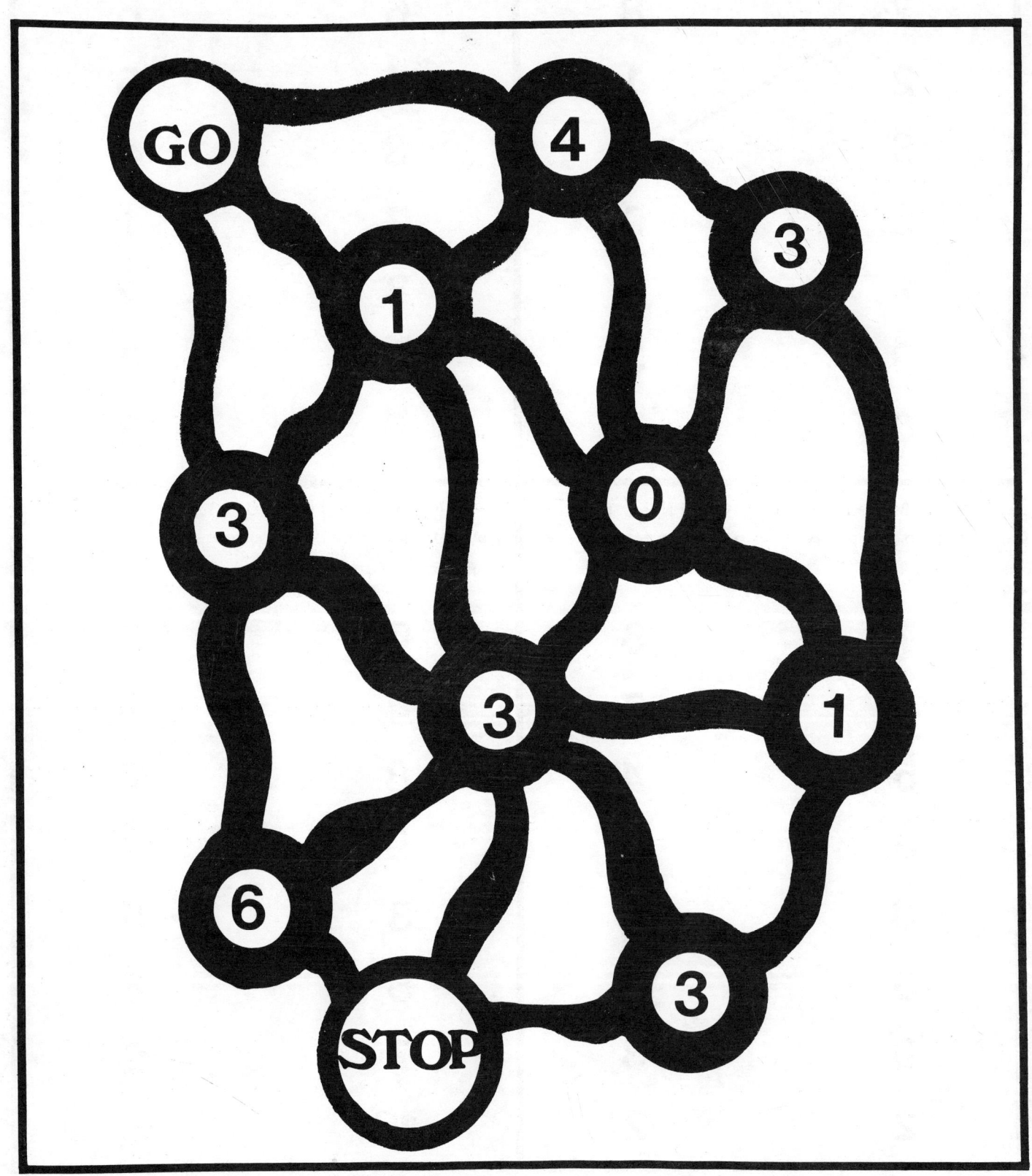

MAGIC SQUARES

4		6
		3
5		

	1	6
7		8

2		
	0	
3	3	

3		
	0	2
		7

MAGIC SQUARES

6	1	
	0	
		9

21

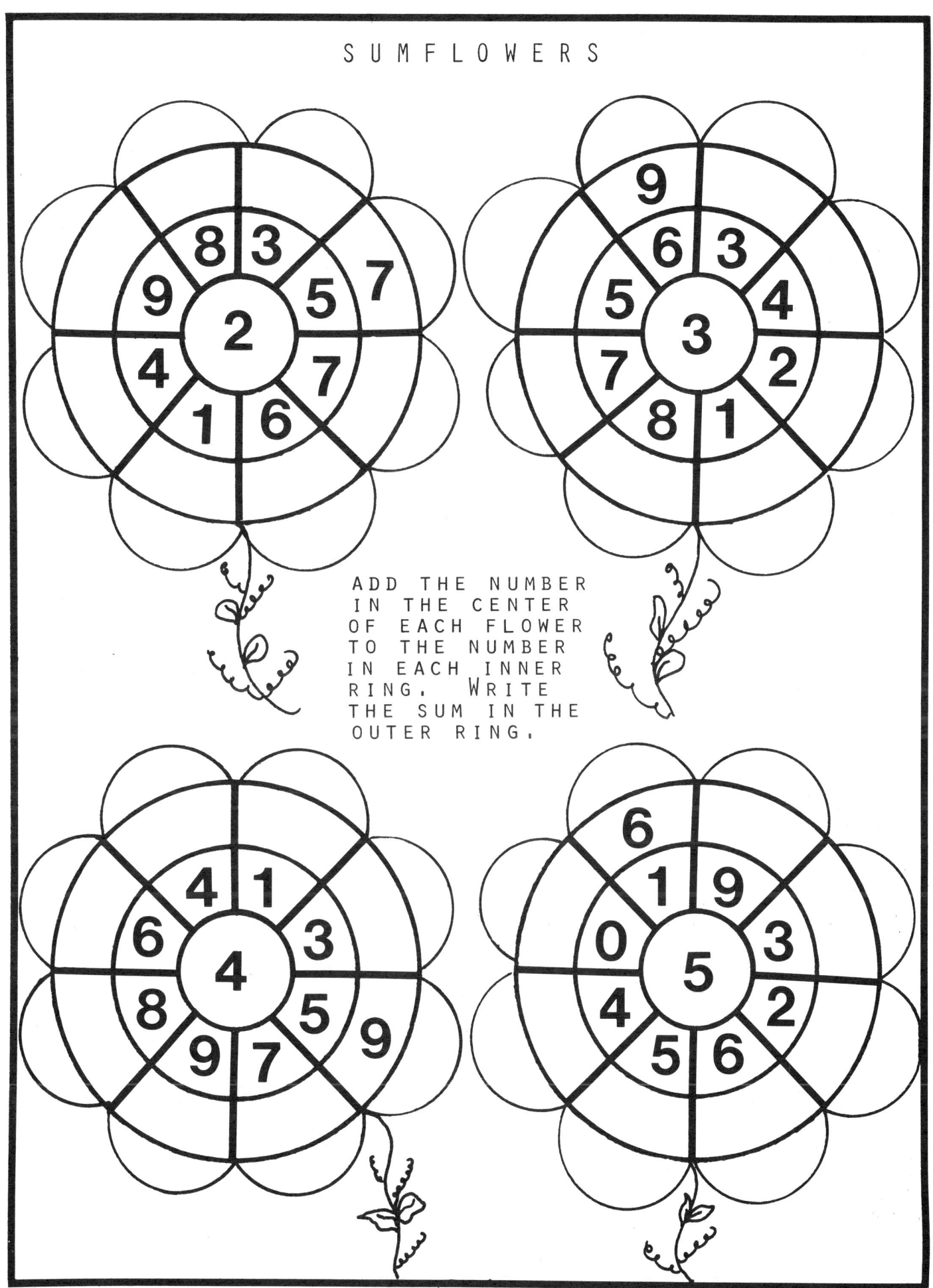

HIDDEN STARS

FIRST SOLVE EACH PROBLEM. THEN COLOR THE PROBLEM AREAS THAT HAVE A SUM OF 10, PINK. COLOR THE PROBLEM AREAS THAT HAVE A SUM OF 11, YELLOW. COLOR THE PROBLEM AREAS WITH ANY OTHER SUM, ORANGE.

I COLORED _____ YELLOW STARS.

I COLORED _____ PINK STARS.

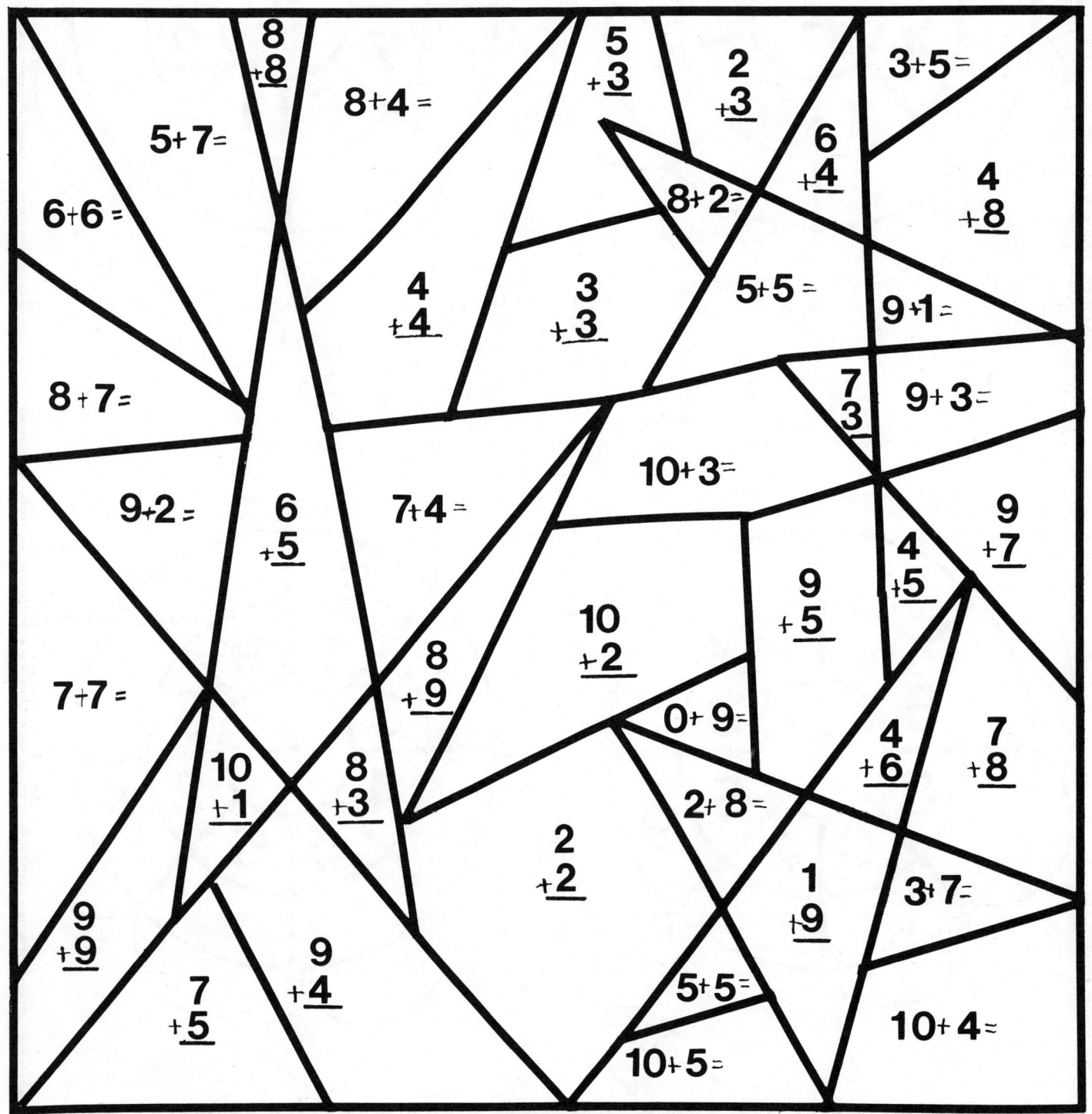

SATELLITE SUMS

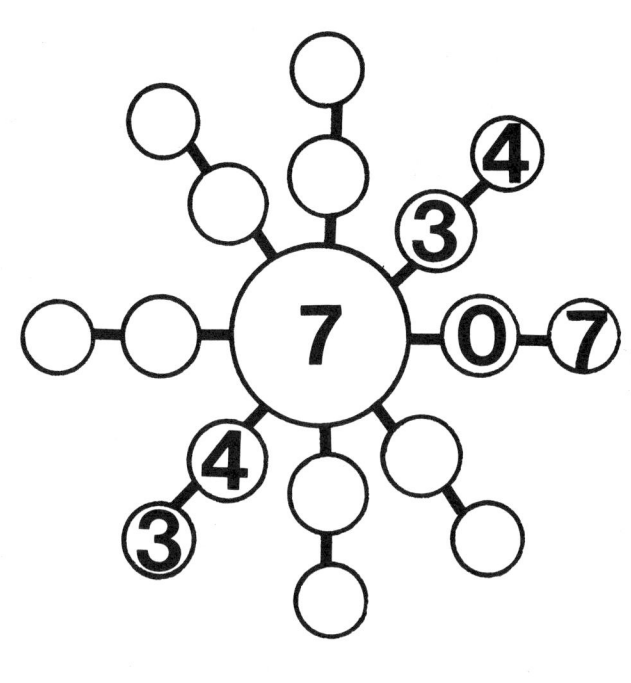

3+4=7
4+3=7
7+0=7

DIRECTIONS: FIND TWO NUMBERS THAT TOTAL THE NUMBER IN THE CENTER OF EACH SATELLITE. WRITE EACH NUMBER PAIR IN A ROW OF THE SMALLER SATELLITE CIRCLES.

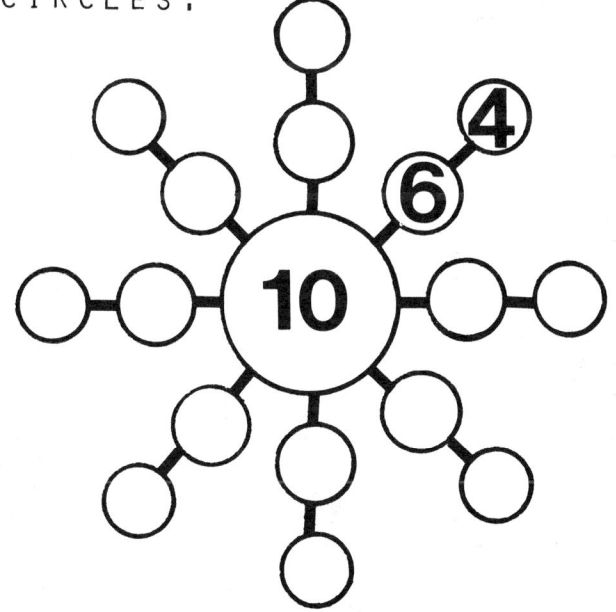
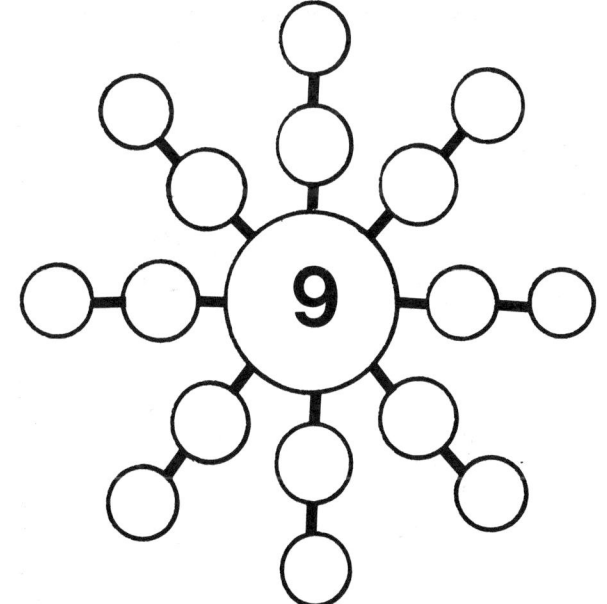

CAN YOU DRAW 2 STRAIGHT LINES THAT WILL DIVIDE EACH SQUARE SO THAT EACH AREA TOTALS 9?

CAN YOU DRAW 2 STRAIGHT LINES THAT WILL DIVIDE EACH SQUARE SO THAT EACH AREA TOTALS 10?

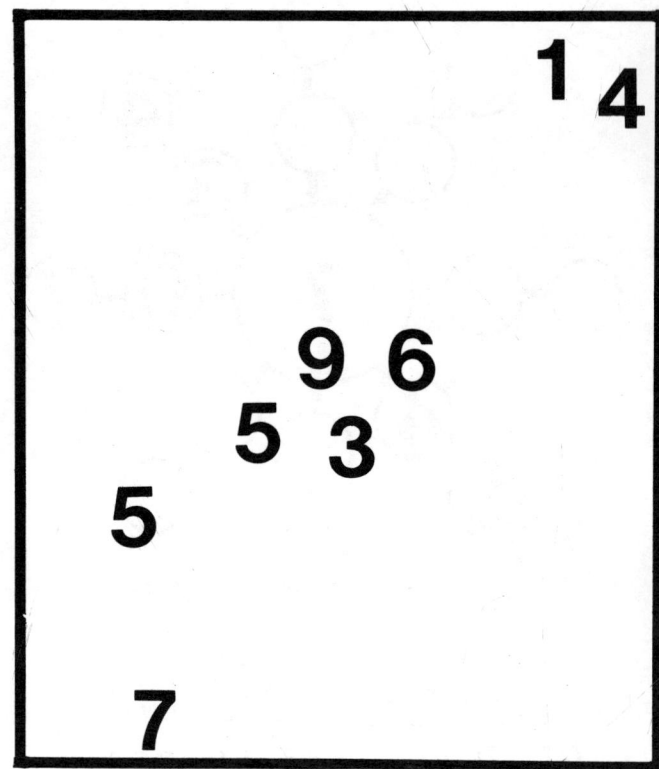

CAN YOU DRAW 2 STRAIGHT LINES THAT WILL DIVIDE EACH SQUARE SO THAT EACH AREA TOTALS 11?

CAN YOU DRAW 2 STRAIGHT LINES THAT WILL DIVIDE EACH SQUARE SO THAT EACH AREA TOTALS 12?

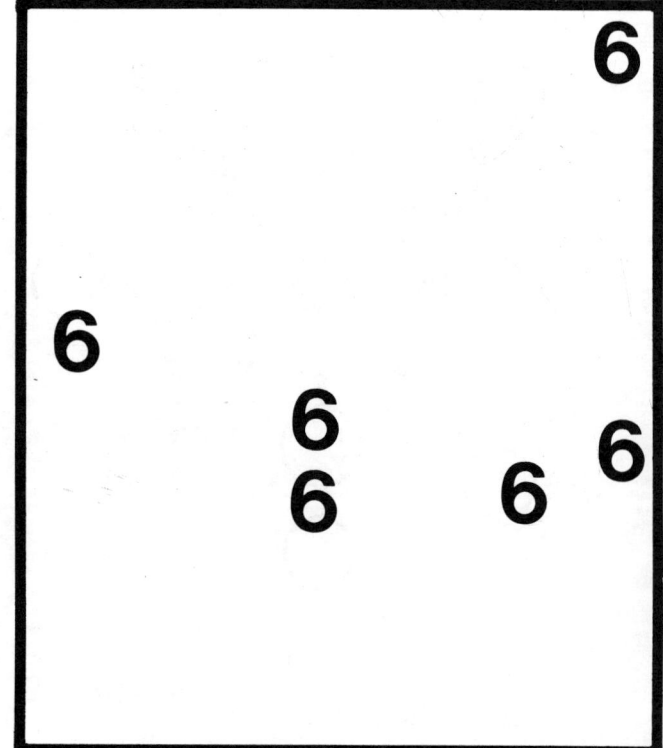

SUM APPLES

Find the sum of all the numbers in each apple. Then answer the following questions:

WHICH APPLES TOTAL MORE THAN 11? _____
WHICH APPLE HAS THE SMALLEST SUM? _____
WHICH APPLE HAS THE SAME SUM AS APPLE "A"?
WHICH APPLE HAS THE SAME SUM AS APPLE "E"?

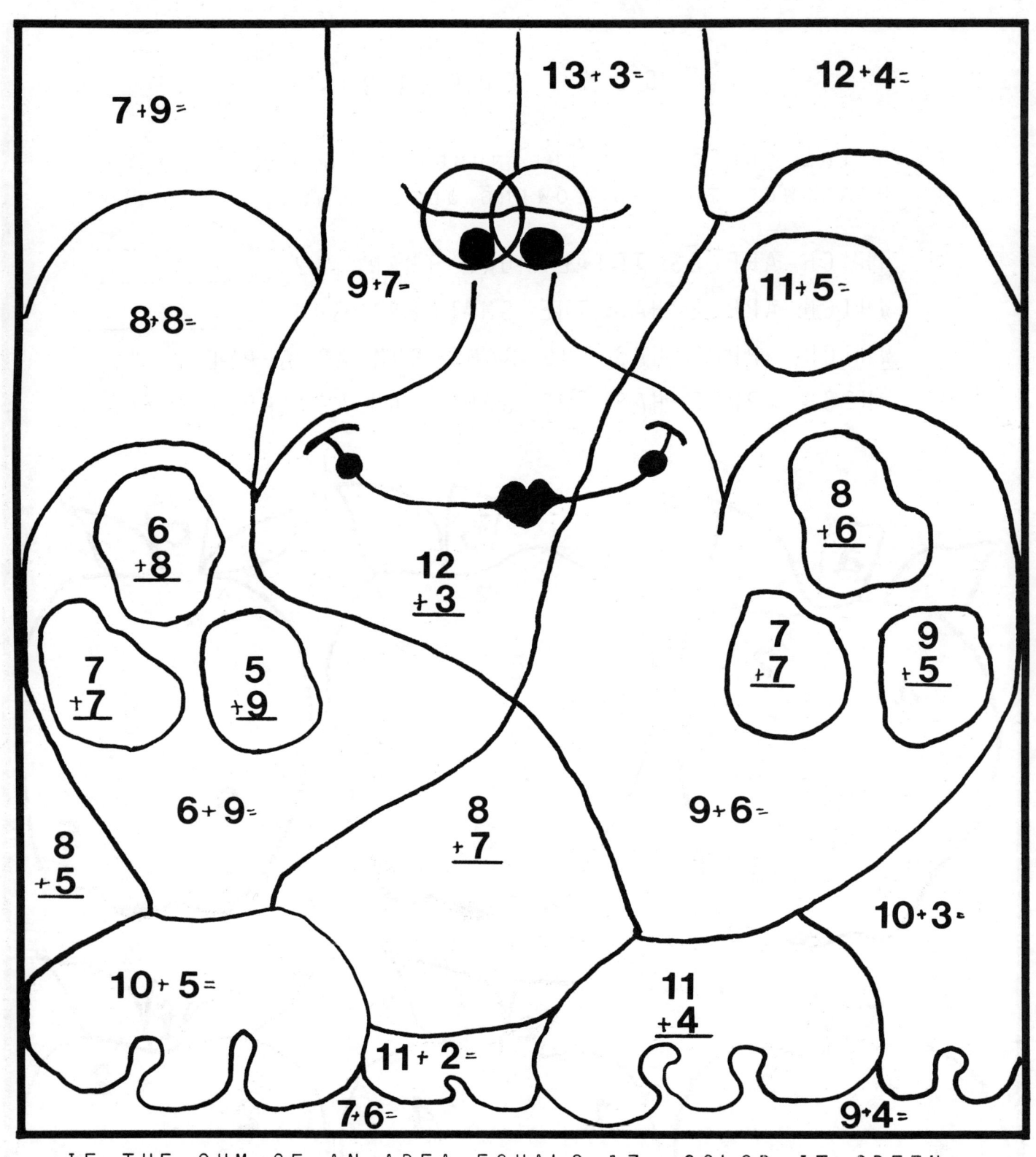

IF THE SUM OF AN AREA EQUALS 13, COLOR IT GREEN
IF THE SUM OF AN AREA EQUALS 14, COLOR IT YELLOW.
IF THE SUM OF AN AREA EQUALS 15, COLOR IT PURPLE.
IF THE SUM OF AN AREA EQUALS 16, COLOR IT PINK.
HOW MANY AREAS ARE: GREEN _____   YELLOW _____
                      PURPLE _____   PINK _____

NUMBER PAIRS

Circle pairs that total 15, red. Circle pairs that total 16, orange.

DRAW A LINE TO CONNECT THE NUMBER PAIRS THAT TOTAL 13.

0	12
1	13
3	9
4	10
2	8
5	11
6	5
8	7

DRAW A LINE TO CONNECT THE NUMBER PAIRS THAT TOTAL 14.

5	9
9	12
2	5
7	10
6	7
4	8
13	11
3	1

DRAW A LINE TO CONNECT THE NUMBER PAIRS THAT TOTAL 15.

10	5
1	6
5	7
9	10
4	2
8	11
3	14
13	12

DRAW A LINE TO CONNECT THE NUMBER PAIRS THAT TOTAL 16.

5	8
9	6
8	13
4	2
3	7
10	11
14	12
15	1

START AT "GO." FINISH AT "STOP." AS YOU PASS THROUGH THE CIRCLES, ADD THE NUMBERS IN THEM TO YOUR TOTAL.

CAN YOU FIND A TRAIL THAT TOTALS 13 ?
CAN YOU FIND A TRAIL THAT TOTALS 14 ?
CAN YOU FIND A TRAIL THAT TOTALS 15 ?
CAN YOU FIND A TRAIL THAT TOTALS 16 ?

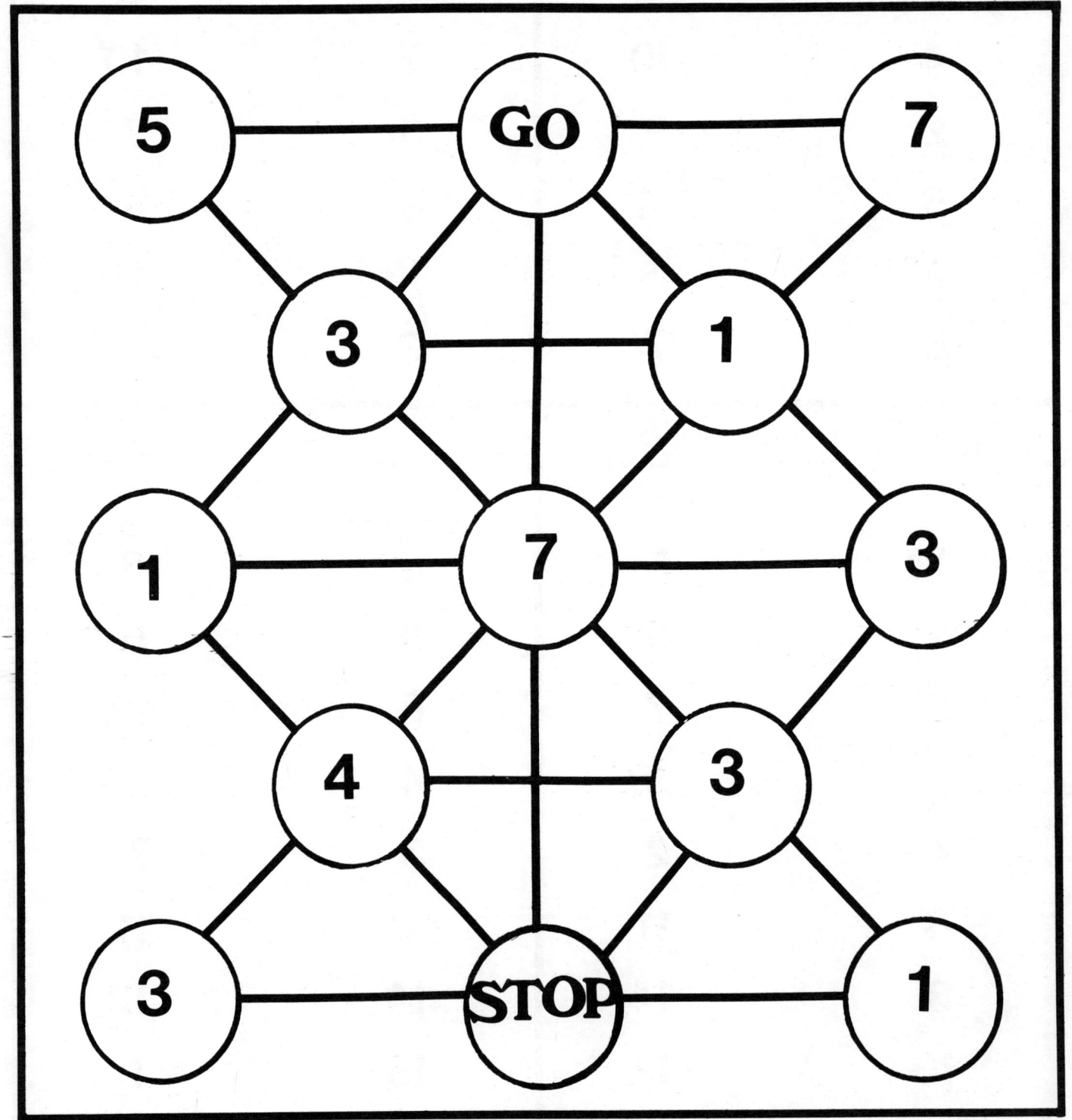

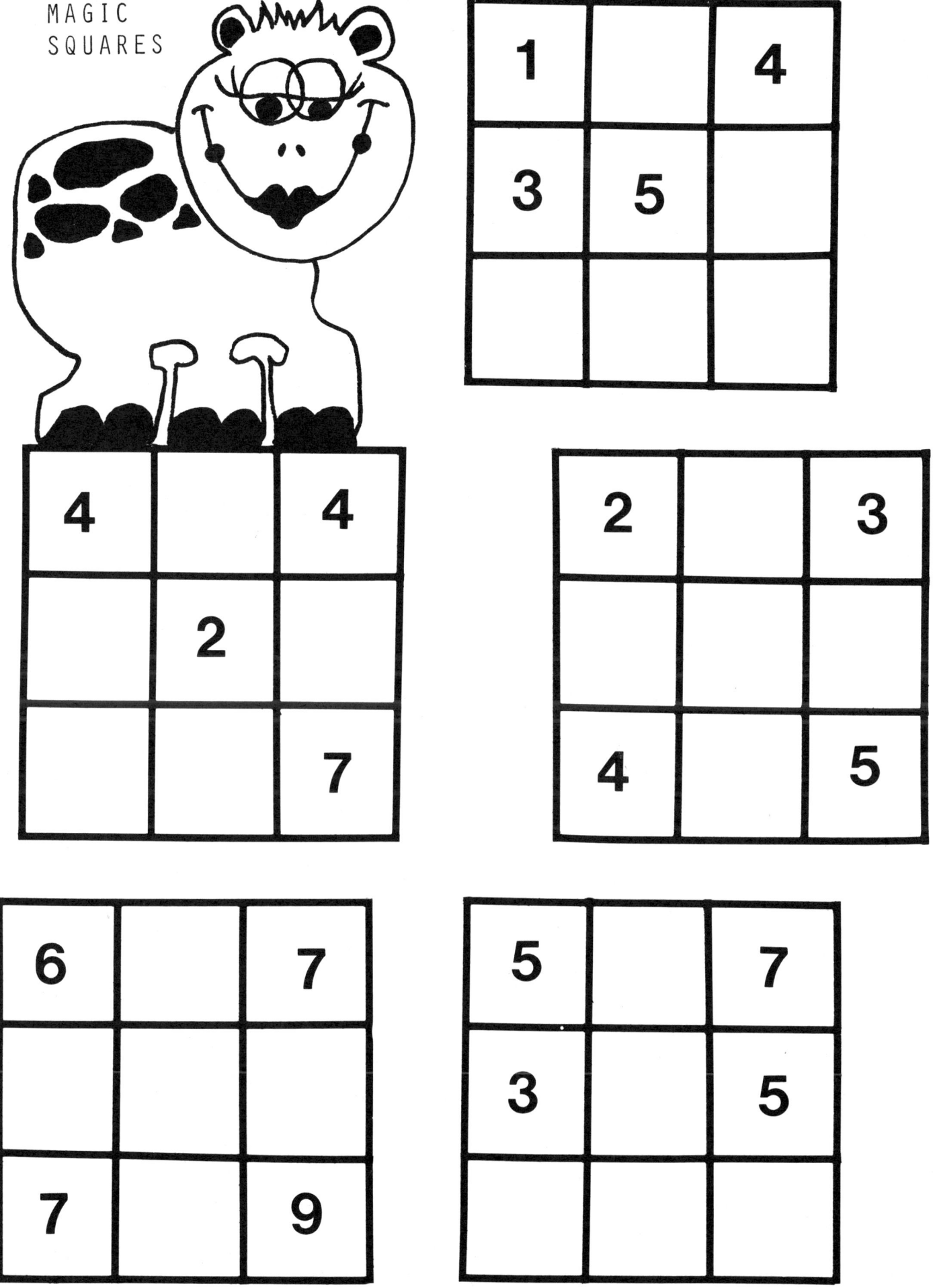

SUMFLOWERS

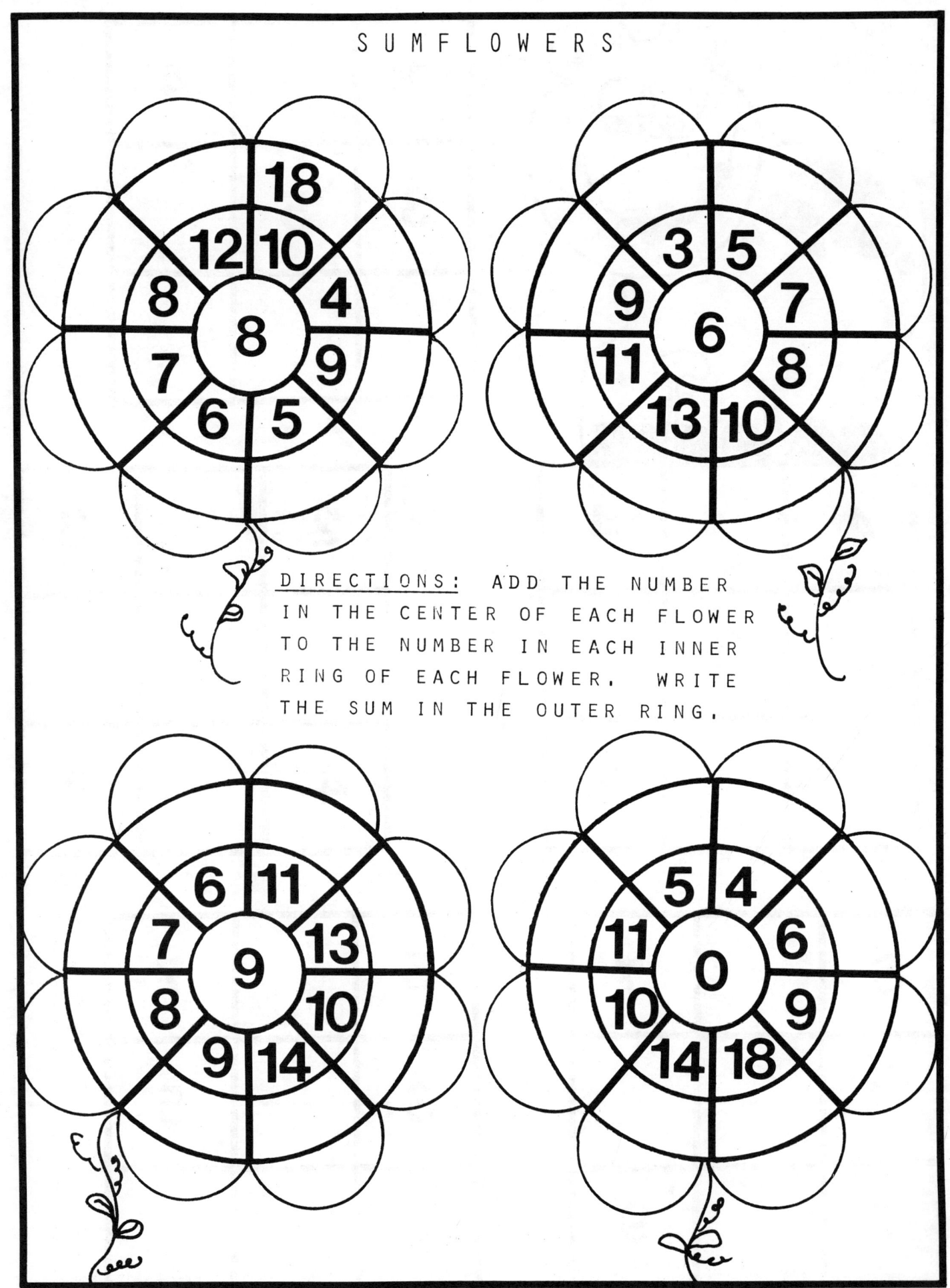

DIRECTIONS: ADD THE NUMBER IN THE CENTER OF EACH FLOWER TO THE NUMBER IN EACH INNER RING OF EACH FLOWER. WRITE THE SUM IN THE OUTER RING.

6 + 6	7 + 9	6 + 10	7 + 6
7 + 7	5 + 7	9 + 6	4 + 8
8 + 8	4 + 9	8 + 6	10 + 4
6 + 7	9 + 5	8 + 5	12 + 4
8 + 7	9 + 6	9 + 3	5 + 10

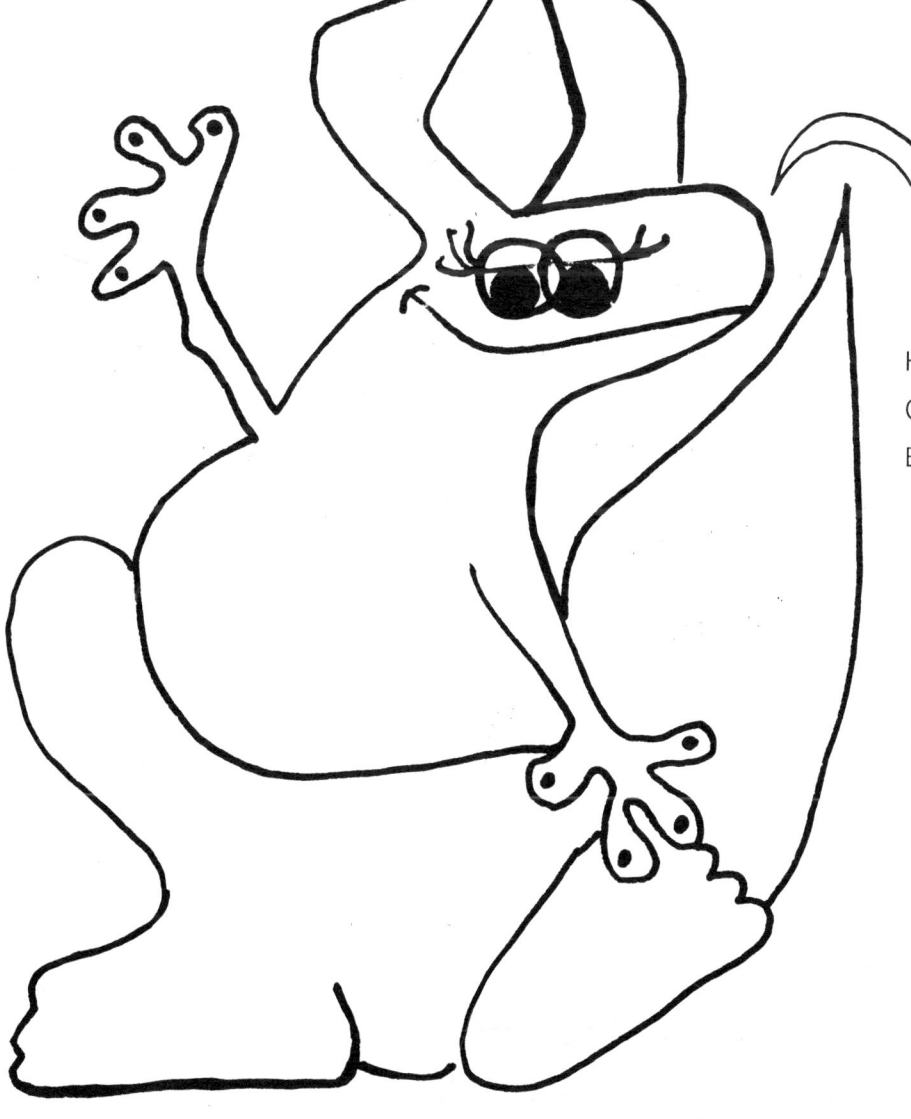

Draw a line to connect the equations that have the same sums.

HOW MANY EQUATIONS CAN YOU WRITE THAT EQUAL 16?

8 + 8 = 16

SATELLITE SUMS

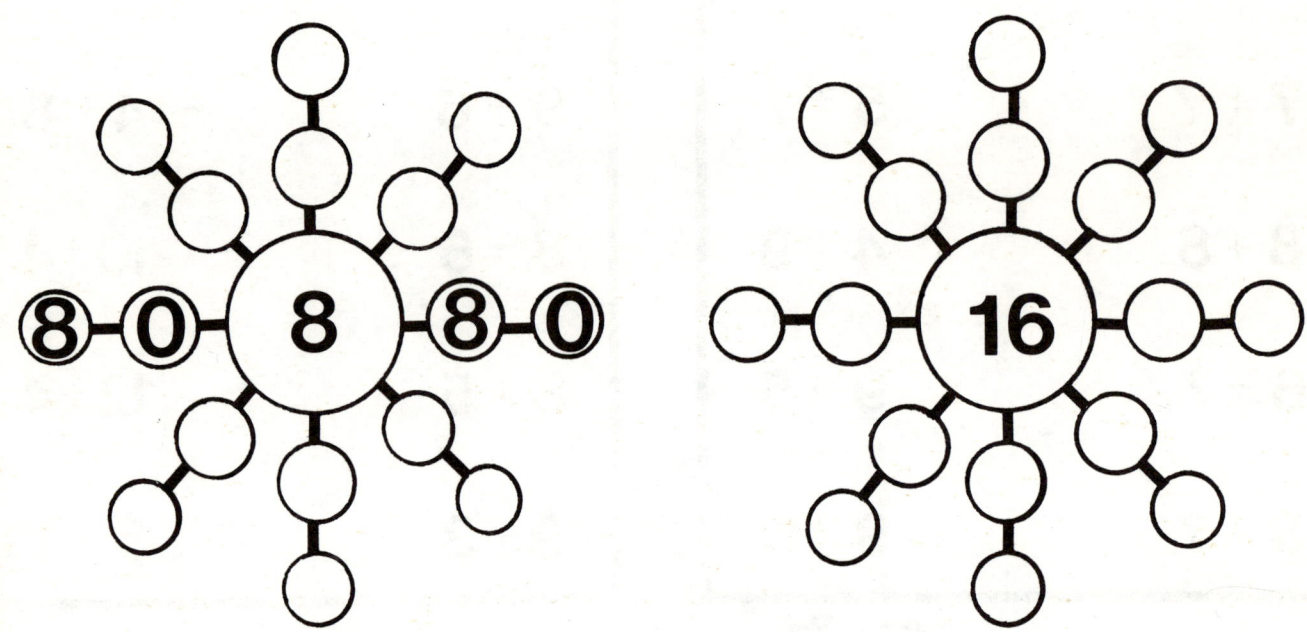

DIRECTIONS: FIND TWO NUMBERS THAT TOTAL THE NUMBER IN THE CENTER OF EACH SATELLITE. WRITE THESE TWO NUMBERS IN THE SMALL SATELLITE CIRCLES.

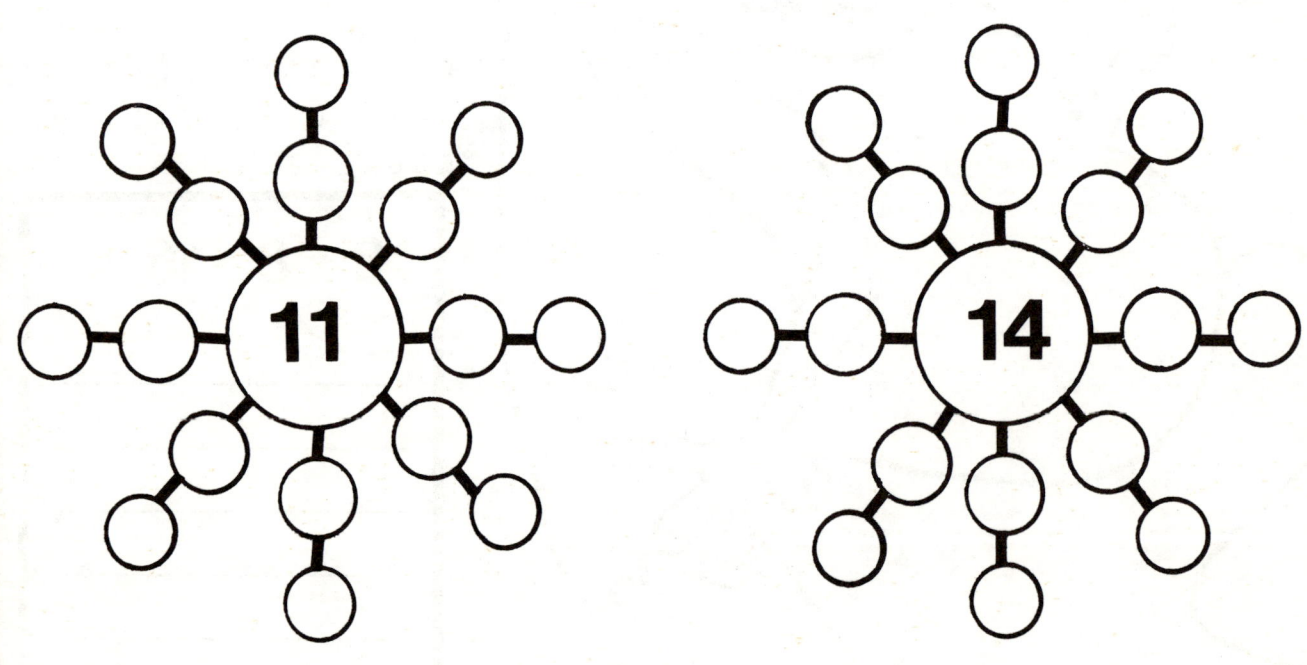

SUPER SATELLITE SUMS

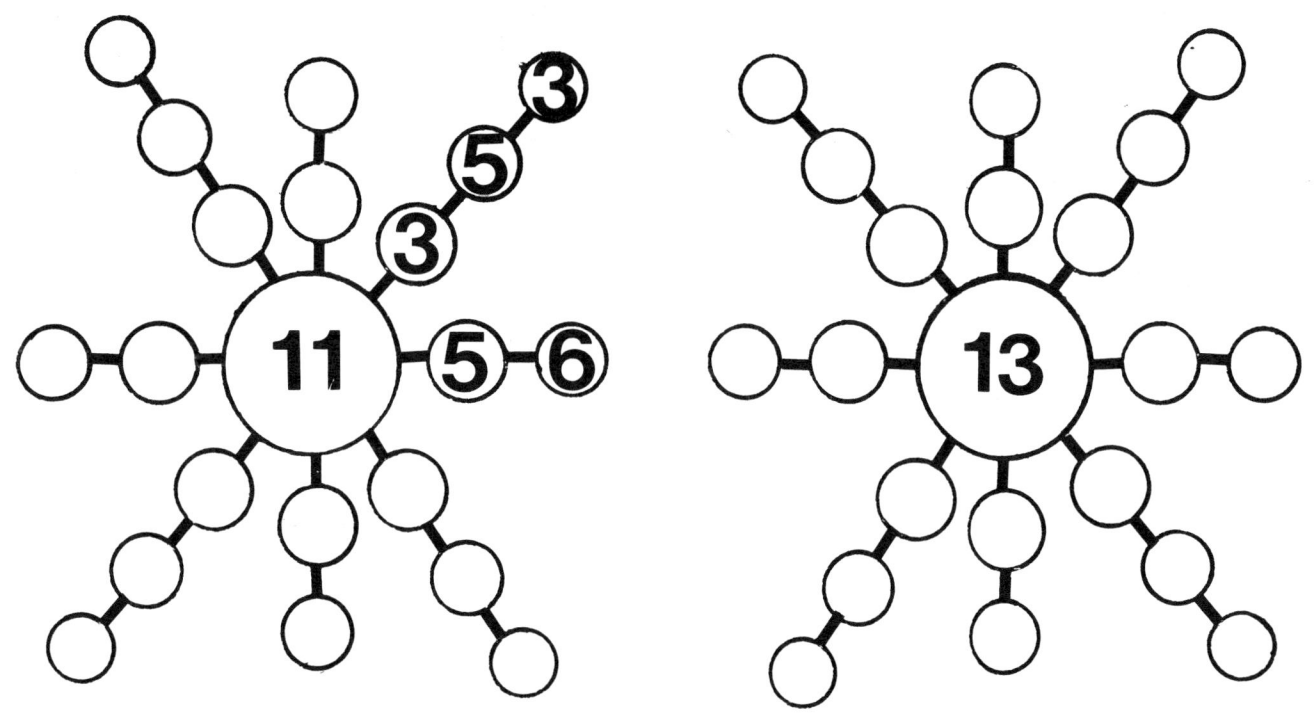

DIRECTIONS: FIND THE TWO OR THREE NUMBERS THAT WHEN TOTALED EQUAL THE NUMBER IN THE CENTER OF THE SATELLITE. WRITE THE ADDENDS IN THE SMALLER CIRCLES OF THE SATELLITE.

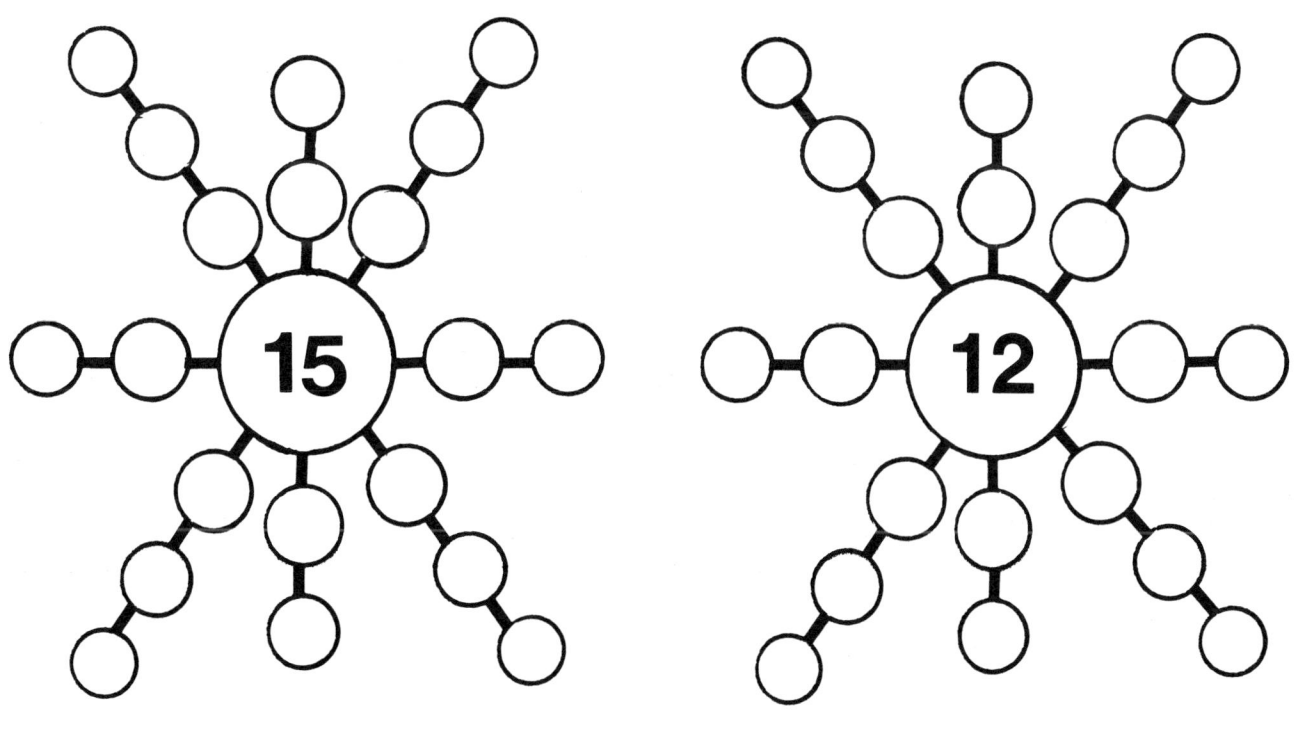

CAN YOU DRAW 2 STRAIGHT LINES THAT WILL DIVIDE THE SQUARE SO THAT EACH AREA TOTALS 13.

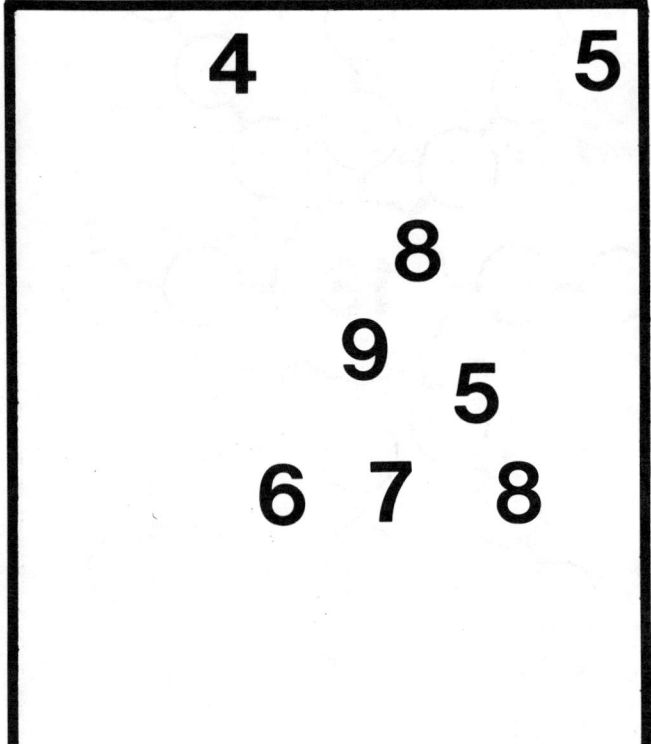

CAN YOU DRAW 2 STRAIGHT LINES THAT WILL DIVIDE THE SQUARE SO THAT EACH AREA TOTALS 14?

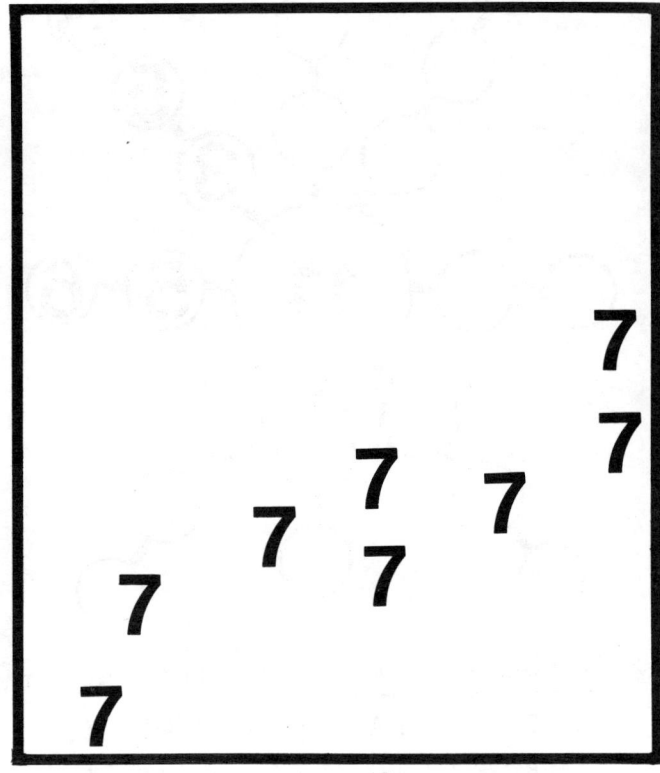

CAN YOU DRAW 2 STRAIGHT LINES THAT WILL DIVIDE THE SQUARE SO THAT EACH AREA TOTALS 15?

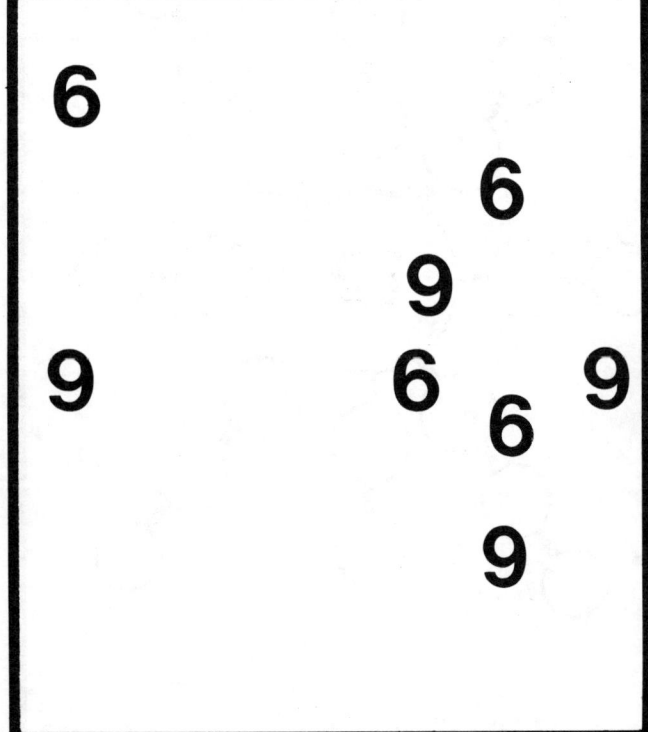

CAN YOU DRAW 2 STRAIGHT LINES THAT WILL DIVIDE THE SQUARE SO THAT EACH AREA TOTALS 16?

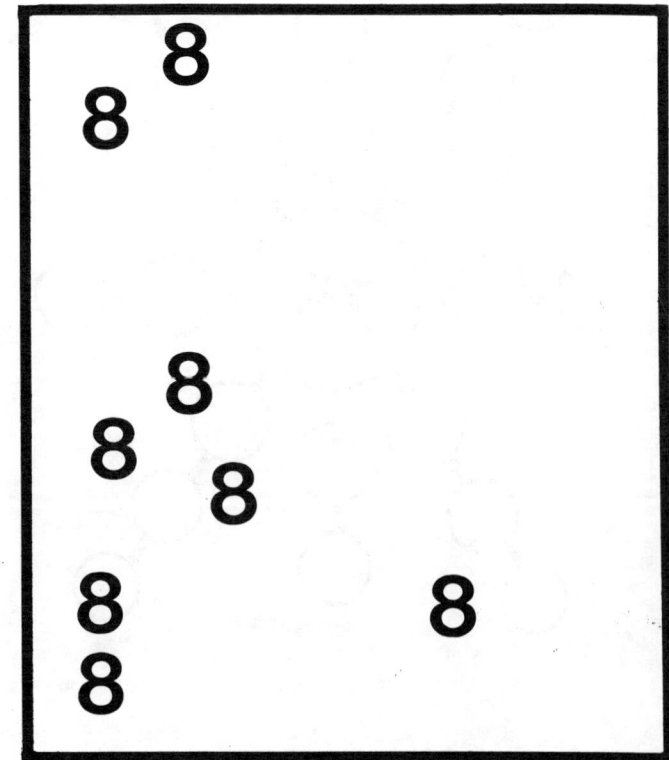

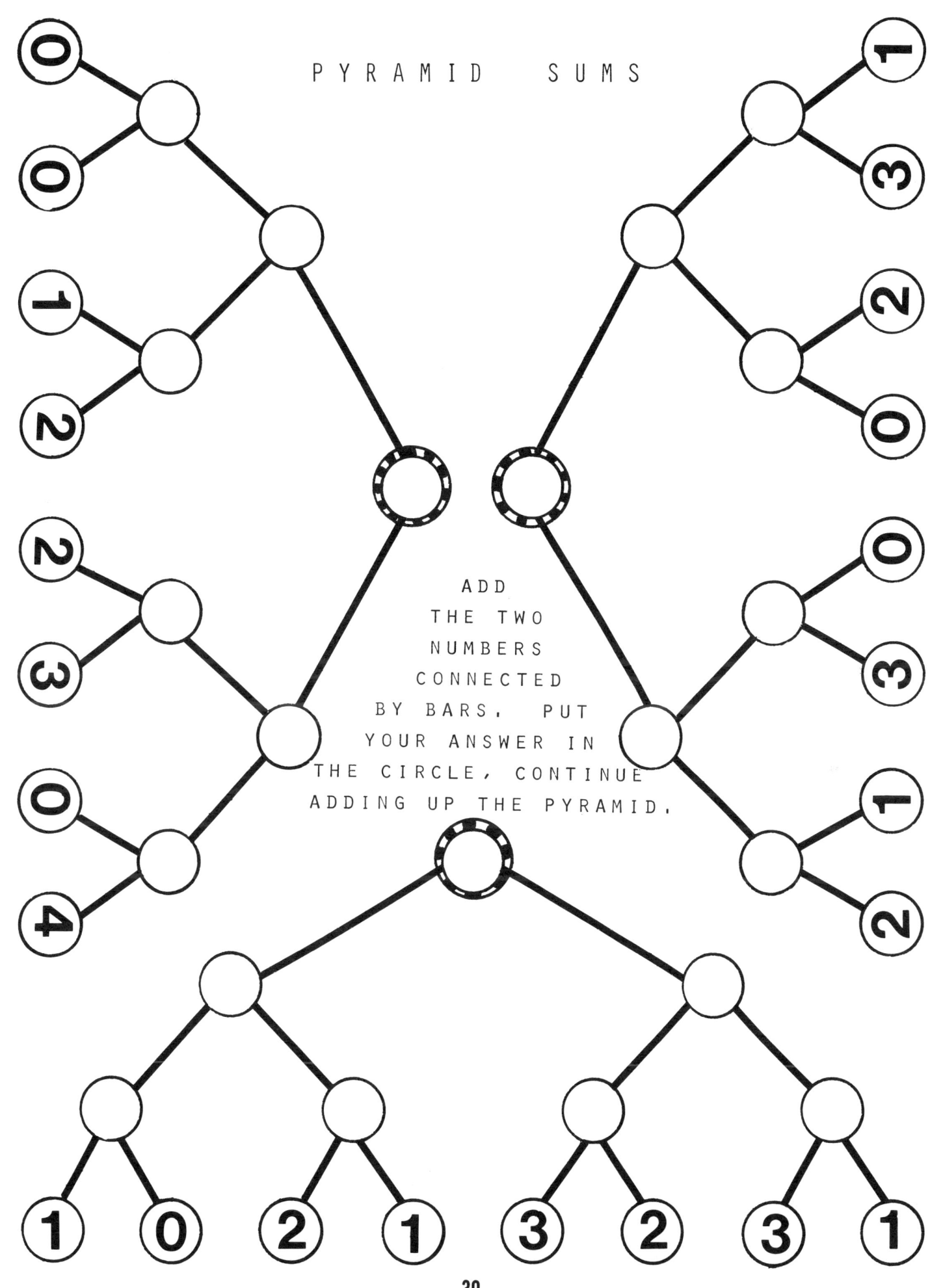

CHART SUMS

1	2	3	4	5	PROBLEMS AND ANSWERS
●		●			1+3=4
	●		●		
●				●	
		●	●		
	●	●			
●			●		
●	●				
	●			●	
		●		●	
			●	●	
●	●	●			1+2+3=6
	●	●		●	
●		●	●		
●	●			●	
●	●		●		
	●	●	●		

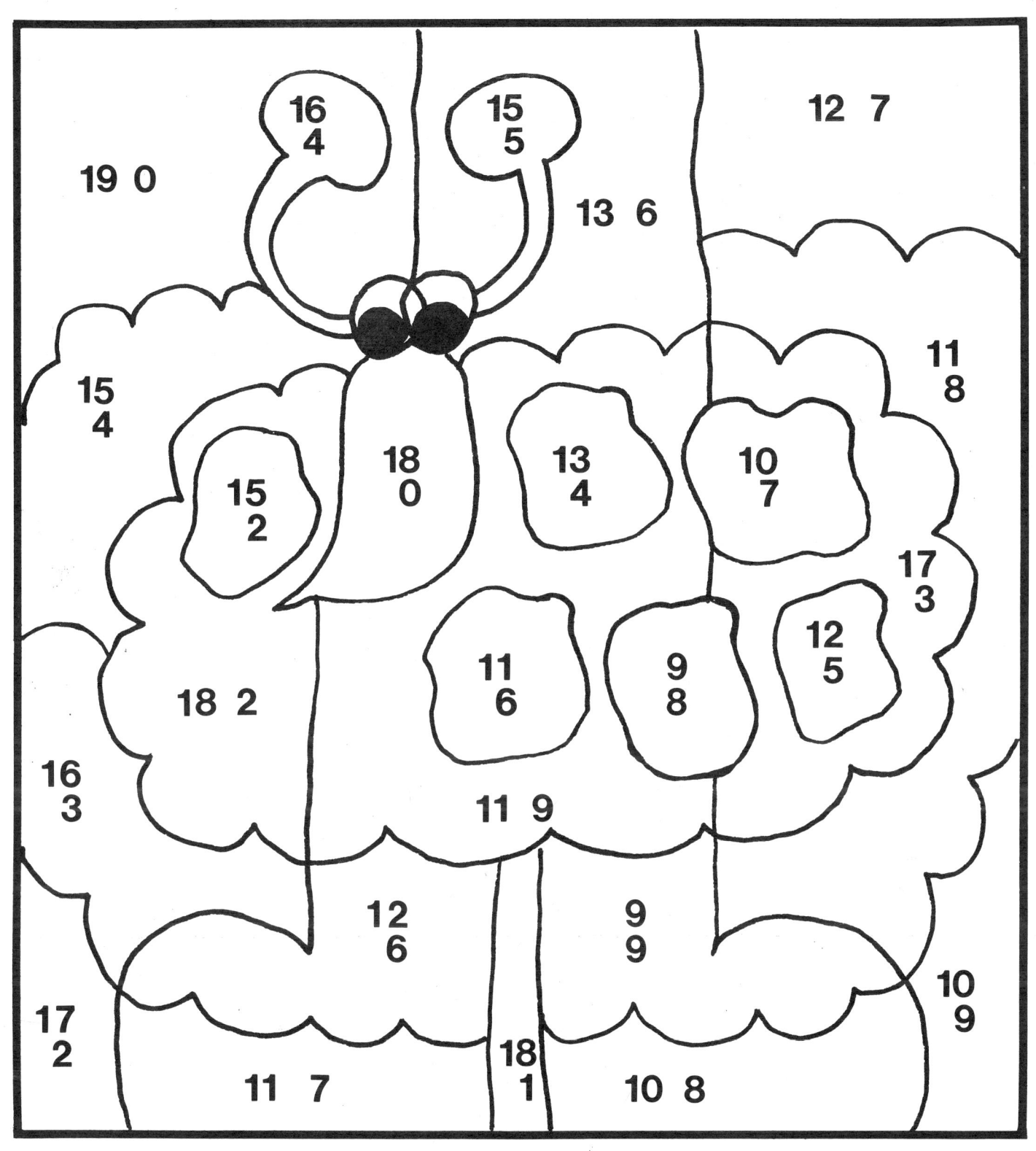

COLOR THE AREAS WHOSE SUMS EQUAL 17, RED.
COLOR THE AREAS WHOSE SUMS EQUAL 18, ORANGE.
COLOR THE AREAS WHOSE SUMS EQUAL 19, PURPLE.
COLOR THE AREAS WHOSE SUMS EQUAL 20, YELLOW.

I COLORED A PICTURE OF A _____.

DRAW A LINE TO CONNECT THE NUMBER PAIRS THAT TOTAL 17.		DRAW A LINE TO CONNECT THE NUMBER PAIRS THAT TOTAL 18.	
7	9	9	9
8	10	10	8
9	8	8	10
10	7	7	11
12	11	11	7
6	5	12	14
13	2	4	5
15	4	13	6

DRAW A LINE TO CONNECT THE NUMBER PAIRS THAT TOTAL 19.		DRAW A LINE TO CONNECT THE NUMBER PAIRS THAT TOTAL 20.	
9	2	10	13
11	10	9	12
17	8	8	10
15	3	7	11
16	4	14	5
14	7	15	6
12	6	13	14
13	5	6	7

CAN YOU FIND A TRAIL THAT TOTALS 17?
CAN YOU FIND A TRAIL THAT TOTALS 18?
CAN YOU FIND A TRAIL THAT TOTALS 19?
CAN YOU FIND A TRAIL THAT TOTALS 21?

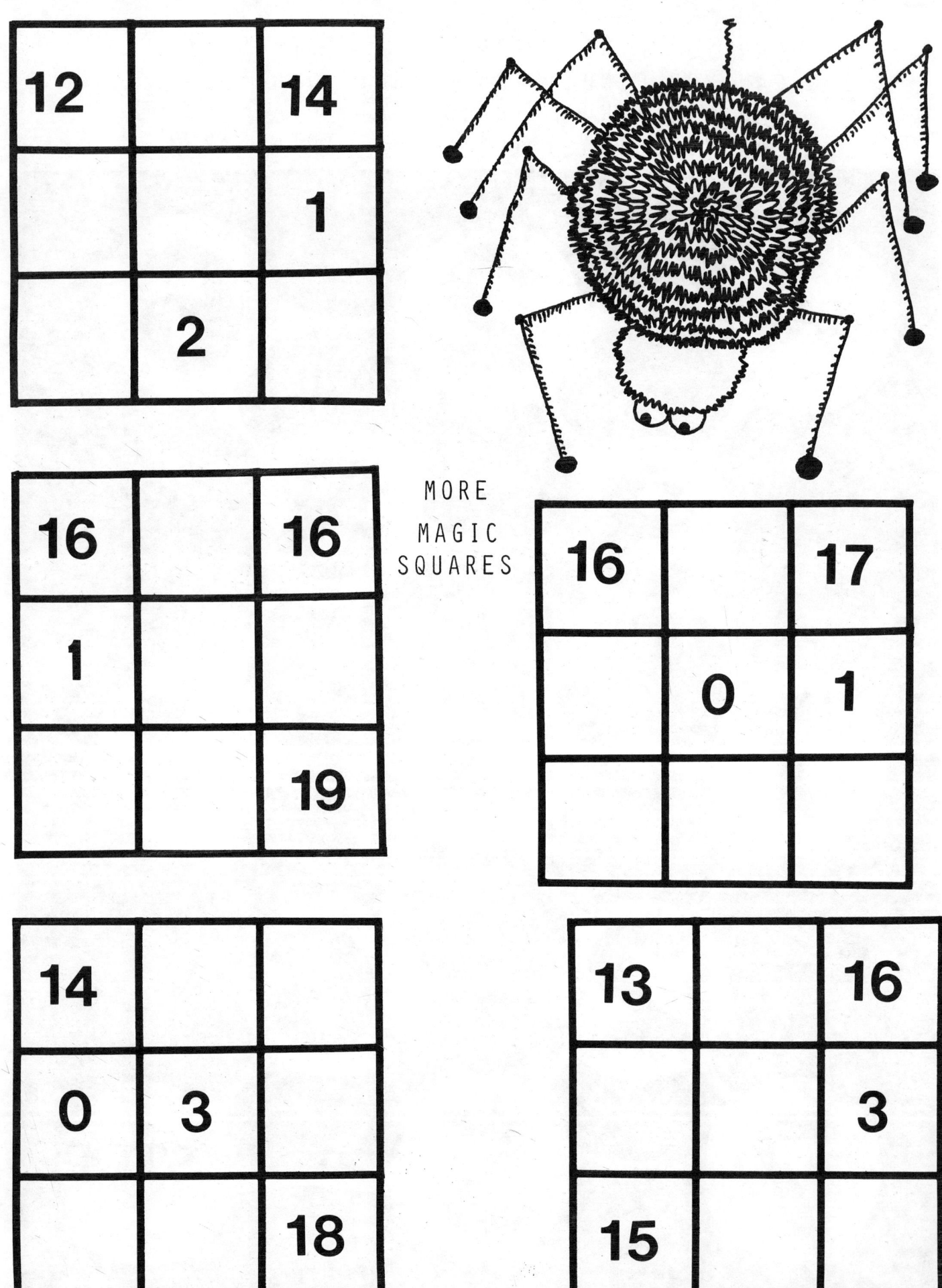

MORE MAGIC SQUARES

MORE MAGIC SQUARES

	4	
		3
3	6	

3		
1	1	
		5

3		
	0	
5		7

	4	9
0		0

3		3
		2
4		

45

MORE CHART SUMS

1	2	3	4	5	6	PROBLEMS AND ANSWERS
●					●	1+6=7
	●				●	
				●	●	
	●	●		●		2+3+5=10
●	●				●	
		●	●		●	
●				●	●	
●		●		●		
●	●			●		
	●	●	●			
●		●			●	
●			●	●		
	●			●	●	
	●	●			●	
			●	●	●	
		●		●	●	
		●	●	●		

MAKE YOUR OWN CHART

<u>DIRECTIONS:</u> MAKE YOUR OWN CHART OF ADDITION SUMS. PUT THREE DOTS IN EACH ROW. THEN WRITE AND SOLVE THE PROBLEMS. DO NOT DO THE SAME PROBLEM MORE THAN ONCE.

1	2	3	4	5	6	PROBLEM AND ANSWER
	•	•		•		2 + 3 = 5

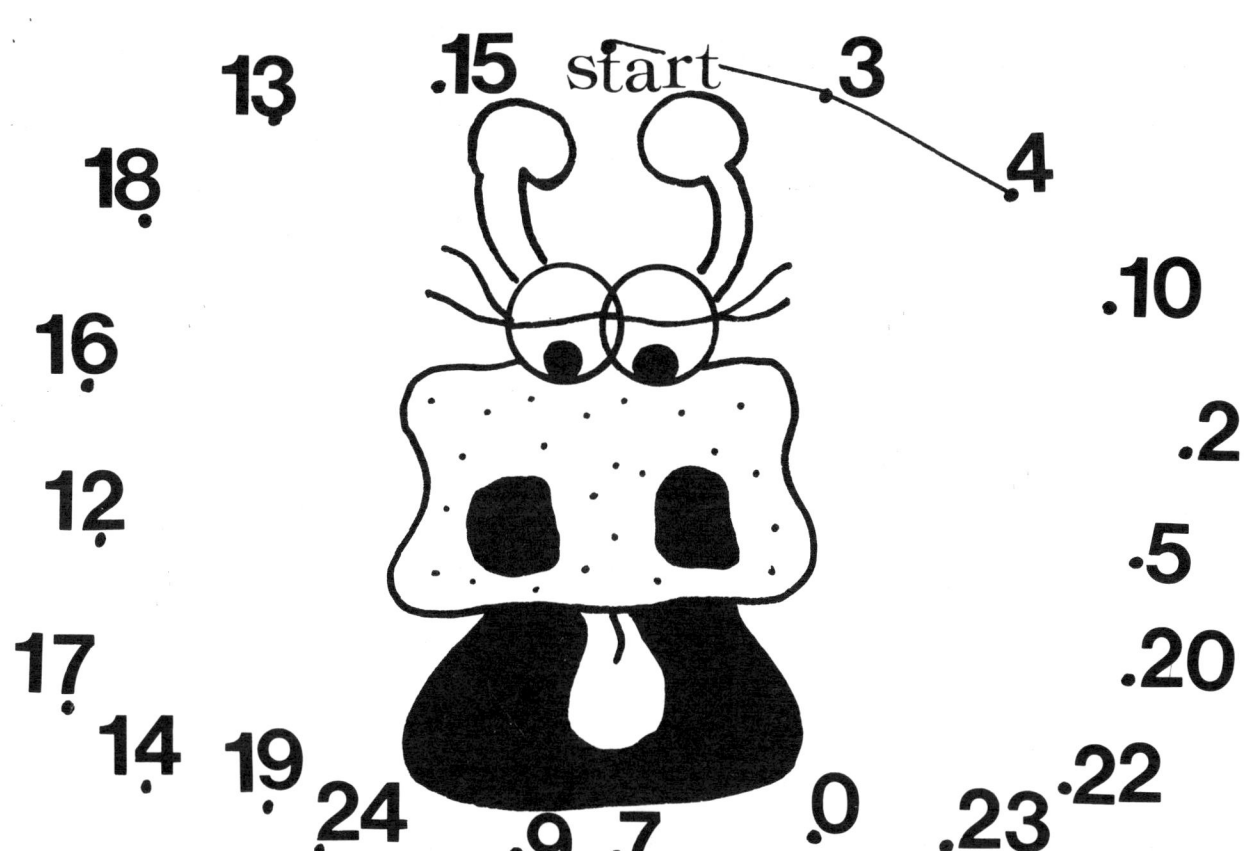

DIRECTIONS: FIRST SOLVE THE PROBLEMS ON THE BOTTOM HALF OF THE PAGE. THEN CONNECT THE ANSWERS IN THE DRAWING AT THE TOP OF THE PAGE.

2 + 1 = 3	0 + 0 =	10 + 9 =
1 + 3 = 4	0 + 1 =	4 + 10 =
5 + 5 =	2 + 4 =	7 + 10 =
1 + 1 =	3 + 4 =	6 + 6 =
3 + 2 =	6 + 3 =	8 + 8 =
10 + 10 =	2 + 6 =	9 + 9 =
11 + 11 =	6 + 5 =	3 + 10 =
10 + 13 =	12 + 12 =	7 + 8 =

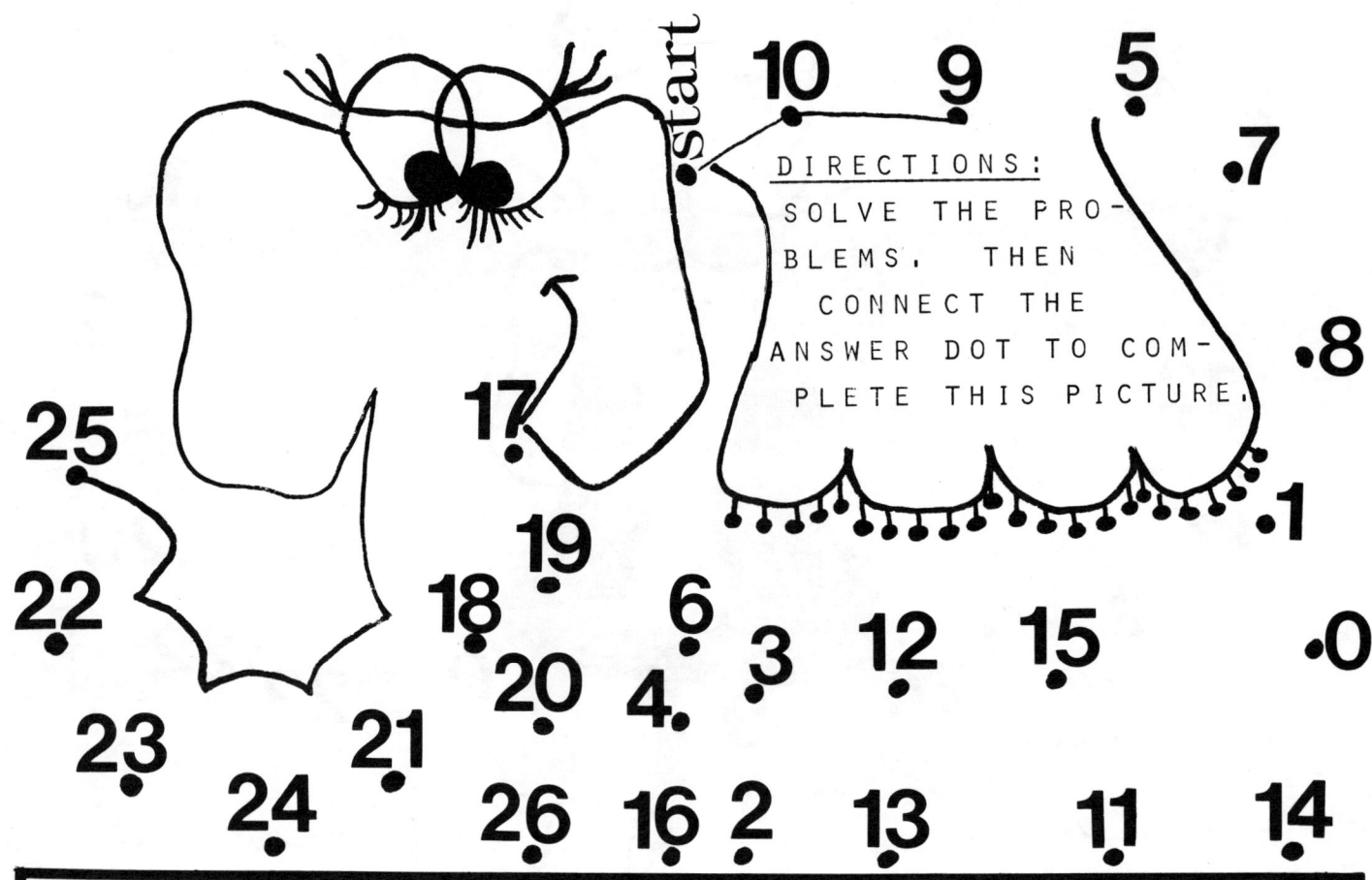

DIRECTIONS: SOLVE THE PROBLEMS. THEN CONNECT THE ANSWER DOT TO COMPLETE THIS PICTURE.

5 + 5 = 10	7 + 8 =	10 + 10 =
3 + 6 = 9	6 + 6 =	10 + 9 =
4 + 1 =	6 + 7 =	10 + 7 =
3 + 4 =	1 + 1 =	10 + 8 =
4 + 4 =	1 + 2 =	20 + 1 =
0 + 1 =	3 + 3 =	20 + 4 =
0 + 0 =	2 + 2 =	20 + 3 =
7 + 7 =	8 + 8 =	20 + 2 =
5 + 6 =	20 + 6 =	20 + 5 =

HIDDEN STARS

FIRST SOLVE ALL THE PROBLEMS. NEXT, COLOR THE AREAS WITH THE SUM OF 19, RED. COLOR THE AREAS WITH THE SUM OF 20, YELLOW. COLOR THE AREAS WITH THE SUM OF ANY OTHER NUMBER, GREEN.

I FOUND _____ RED STARS.

I FOUND _____ YELLOW STARS.

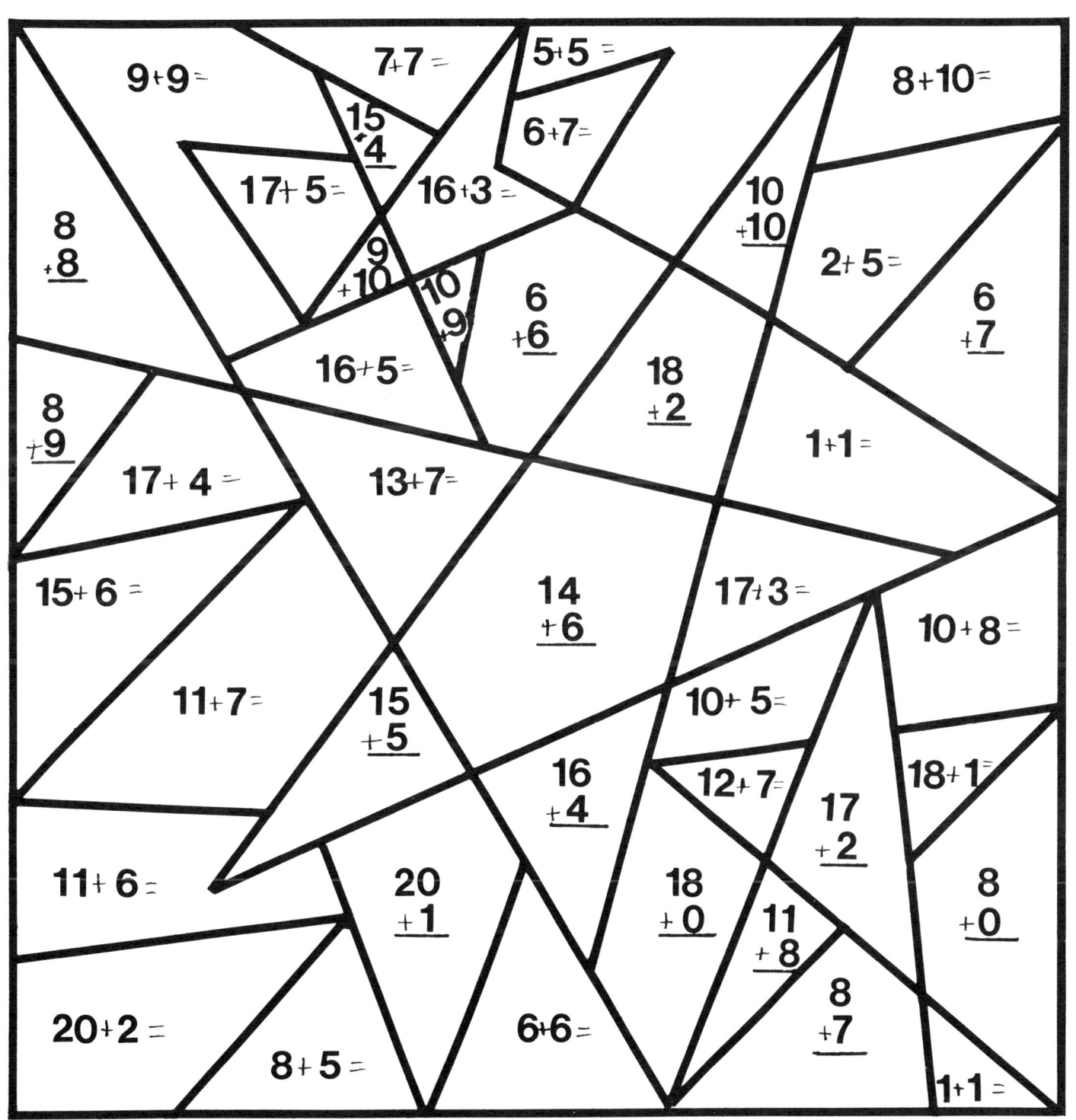

SUPER SATELLITE SUMS

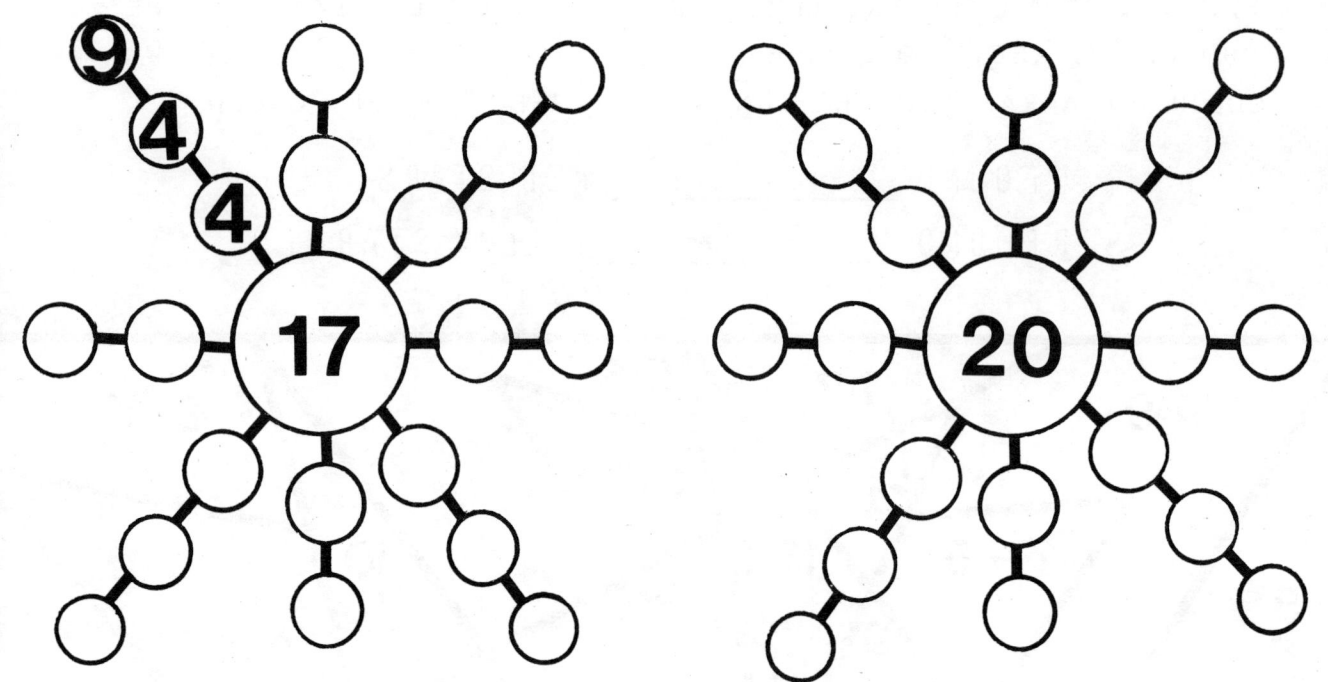

DIRECTIONS: FIND TWO OR THREE NUMBERS THAT WHEN TOTALED EQUAL THE NUMBER IN THE CENTER OF THE SATELLITE. WRITE THE ADDENDS IN THE SMALLER CIRCLES OF EACH SATELLITE.

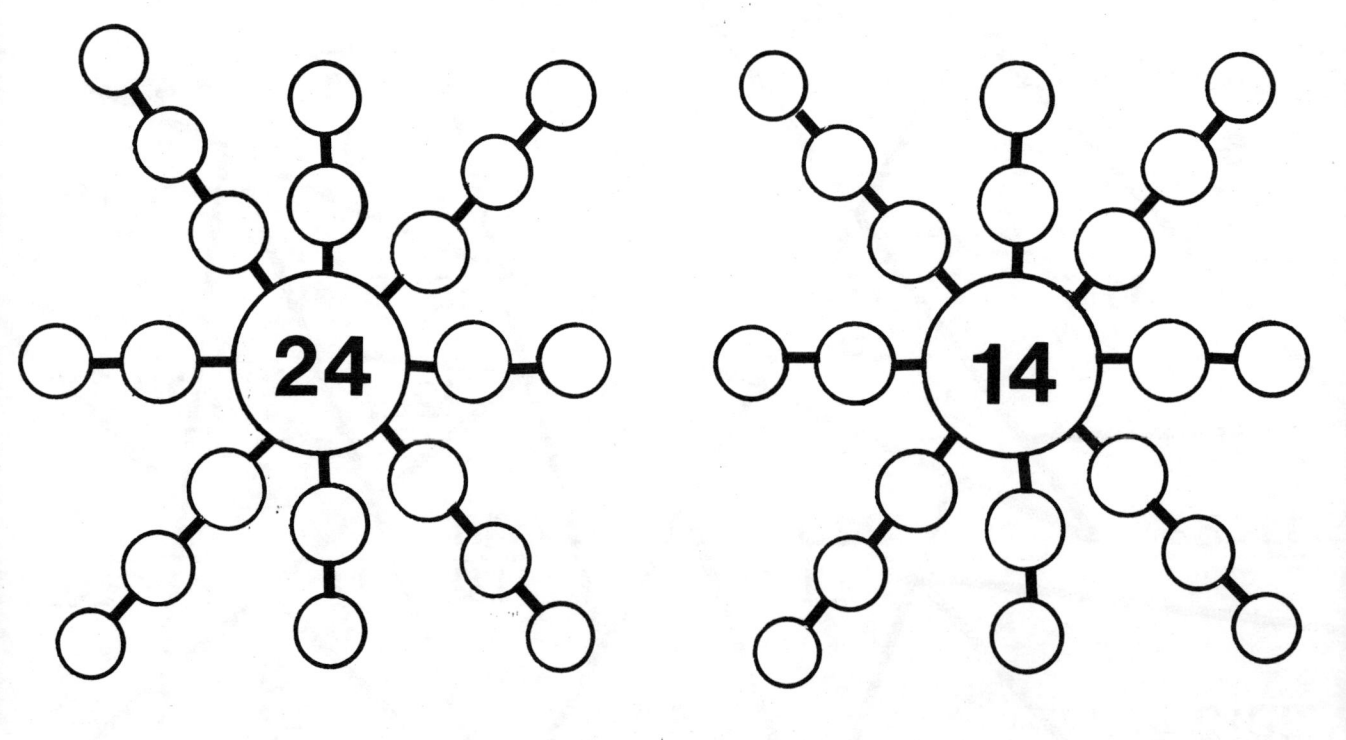

DIRECTIONS: ADD THE NUMBERS FOUND ALONG EACH PATH. THEN ANSWER THE FOLLOWING QUESTIONS:

1. HOW MUCH DOES PATH "Y" EQUAL? _____

2. WHICH PATH TOTALS 18? _____

3. WHICH PATH TOTALS THE MOST? _____

4. WHICH PATH TOTALS AN ODD SUM? _____

5. WHAT IS THE TOTAL OF ALL THREE PATHS? _____

4 + 4 = 8

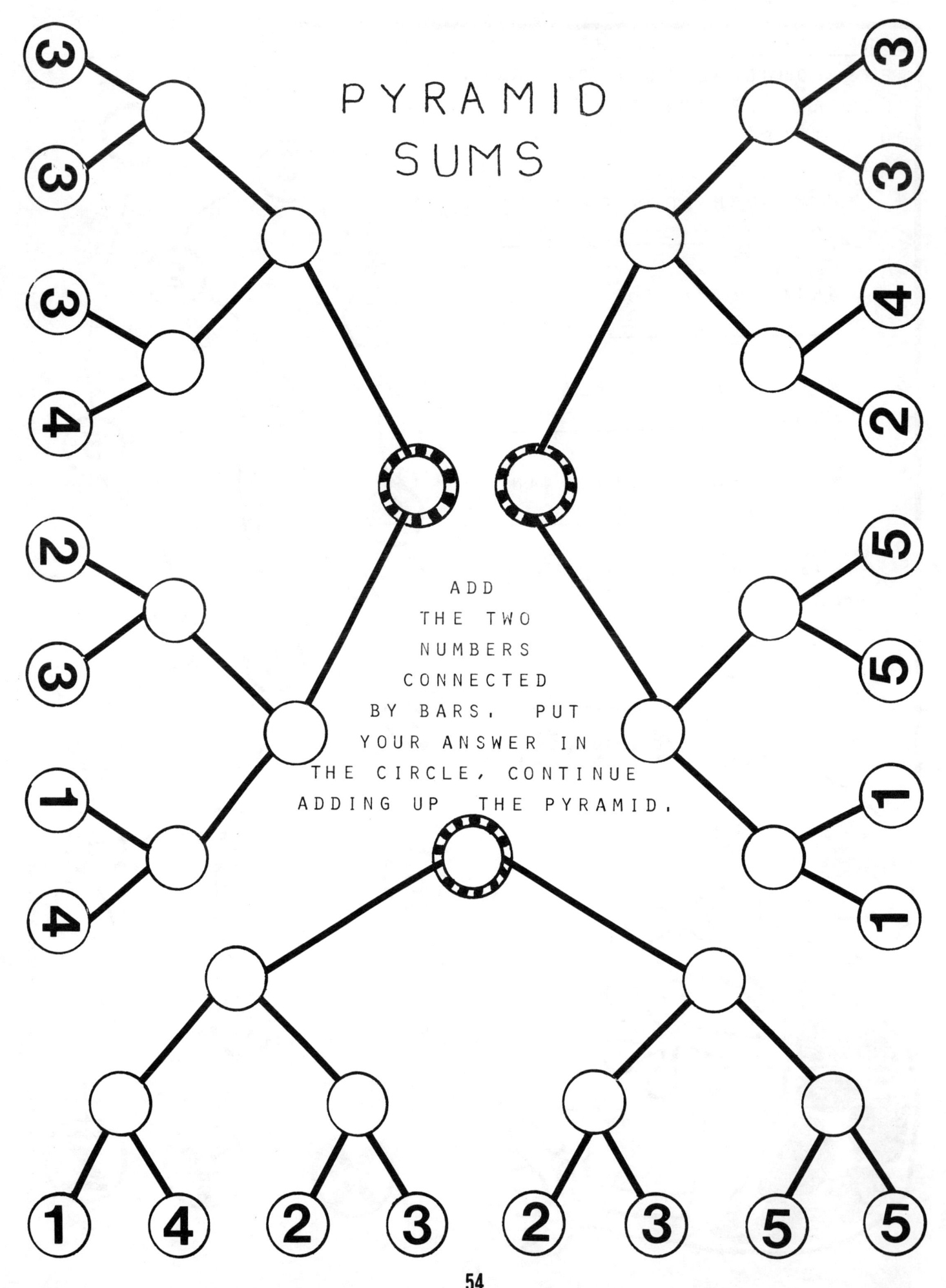

```
A 1      N 14
B 2      O 15
C 3      P 16
D 4      Q 17
E 5      R 18
F 6      S 19
G 7      T 20
H 8      U 21
I 9      V 22
J 10     W 23
K 11     X 24
L 12     Y 25
M 13     Z 26
```

DIRECTIONS: EACH LETTER IS WORTH THE NUMBER OF POINTS LISTED NEXT TO IT.

HOW MUCH ARE THESE TWO-LETTER WORDS WORTH?

IS   9 + 19 = 28

IF   _____

TO   _____

ME   _____

BE   _____

MY   _____

AS   _____

IN   _____

SO   _____

HOW MUCH IS YOUR FIRST NAME WORTH?

_____

WHICH IS WORTH MORE, COOKIES OR CANDY? _____

WHICH IS WORTH MORE, FIRE OR ICE? _____

WHICH IS WORTH MORE, DOGS, OR CATS? _____

WHICH IS WORTH MORE, HATS OR COATS? _____

NOW CAN YOU THINK OF SOME OTHERS?

_____

DIRECTIONS:
ADD THE THREE NUMBERS IN EACH SPACE. IF THE SUM IS 20, COLOR THE SPACE GREEN. IF THE SUM IS 21, COLOR THE SPACE ORANGE. IF THE SUM IS 22, COLOR THE SPACE PINK. IF THE SUM IS 23, COLOR THE SPACE YELLOW.

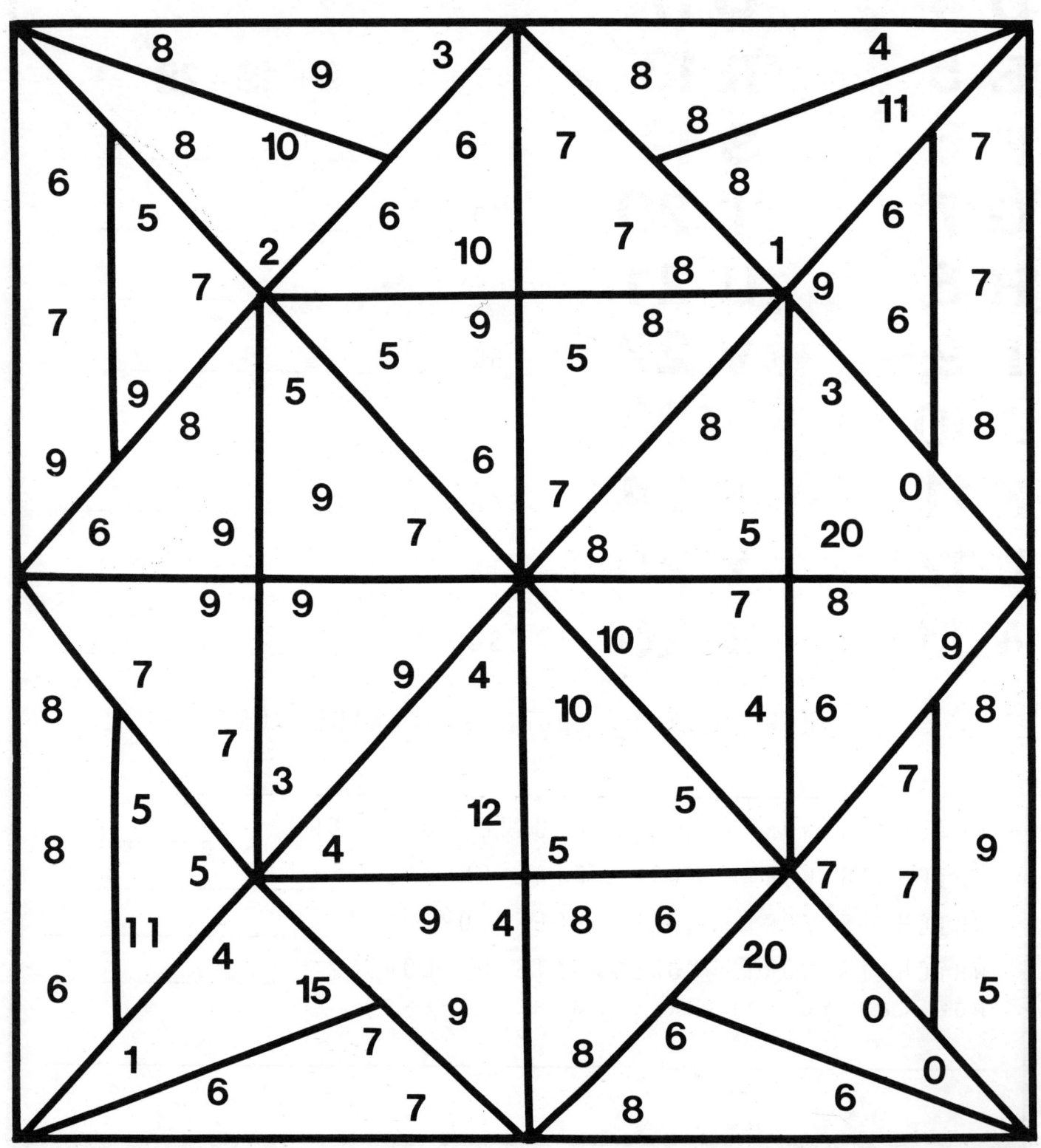

WORK AROUND THE CIRCLE. ADD THE NUMBER IN THE FIRST CIRCLE TO THE LITTLE NUMBER. WRITE THE SUM IN THE NEXT CIRCLE. CONTINUE UNTIL THE CIRCLE IS COMPLETED.

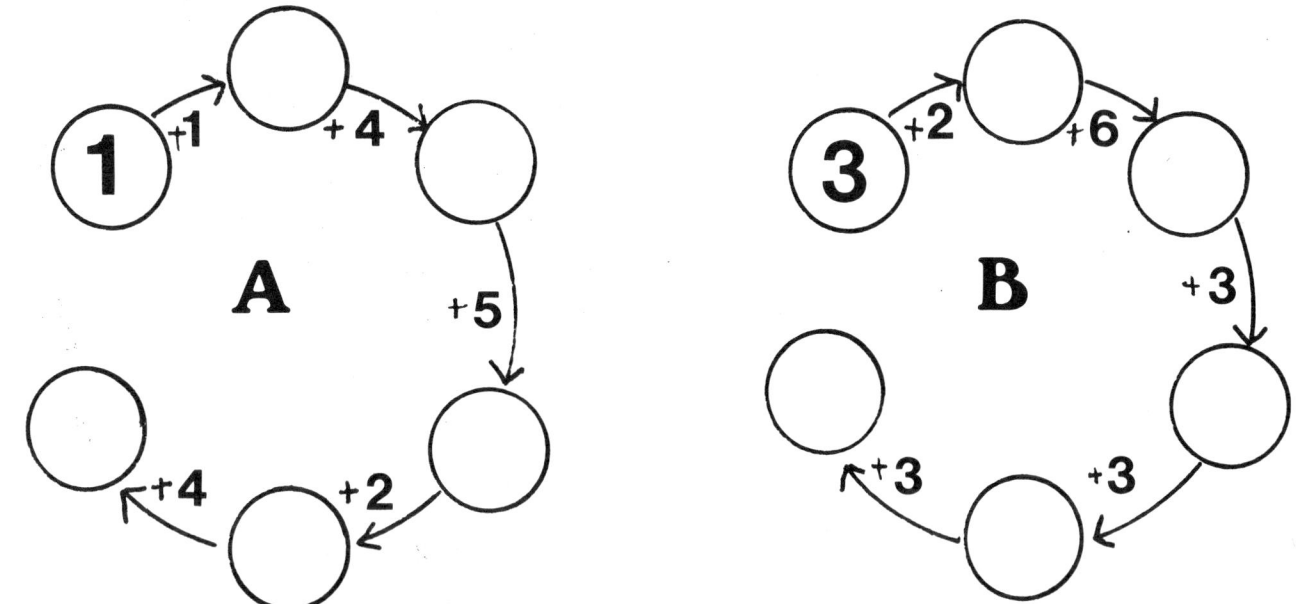

WHICH CIRCLE HAS THE GREATEST SUM? A, B, C, OR D
WHICH CIRCLE HAS THE SMALLEST SUM? A, B, C, OR D
WHICH CIRCLES HAVE AN EVEN NUMBER SUM? A, B, C, OR D
WHICH CIRCLES HAVE AN ODD NUMBER SUM? A, B, C, OR D

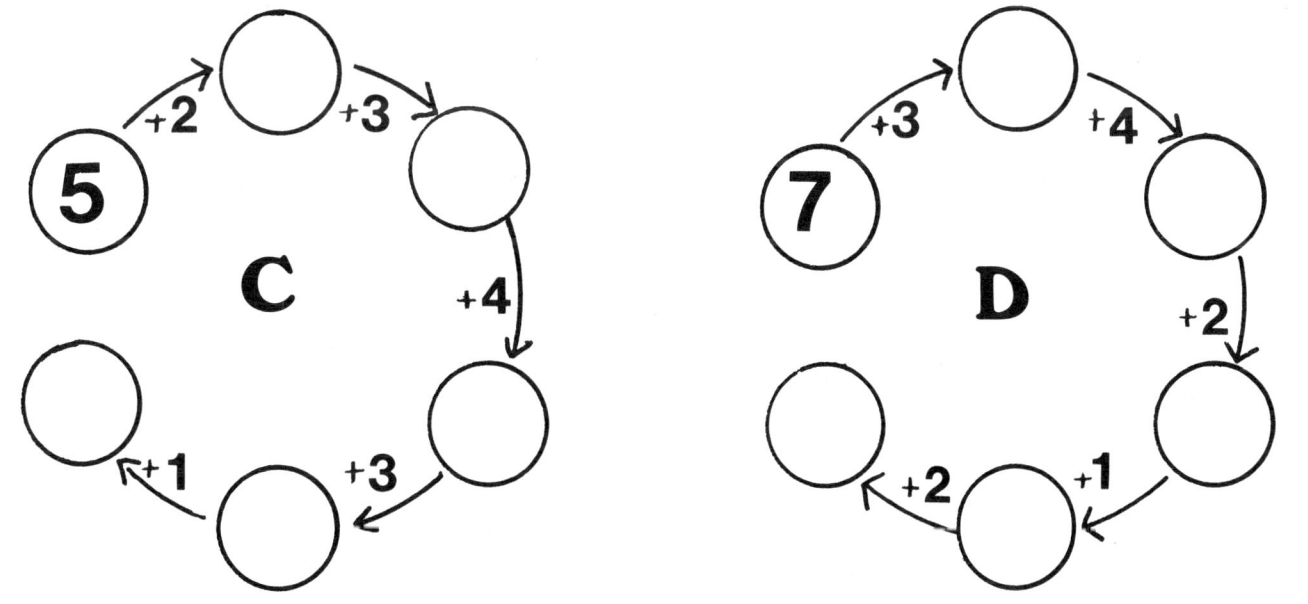

WHICH CIRCLE HAS A SUM ONE LESS THAN CIRCLE B? _____
WHICH CIRCLE HAS A SUM TWO LESS THAN CIRCLE D? _____

JUST FOR FUN

CHOOSE A NUMBER. ADD THE TOP NUMBER IN EACH ROW THAT YOUR NUMBER APPEARS IN.

FOR EXAMPLE: NUMBER 11. NUMBER 11 IS IN ROWS MARKED 1, 2, AND 8. SO, 1 + 2 + 8 = 11.

1	2	4	8
1	2	4	8
3	3	5	9
5	6	6	10
7	7	7	11
9	10	12	12
11	11	13	13
13	14	14	14
15	15	15	15

ANOTHER EXAMPLE WOULD BE THE NUMBER 7. SEVEN IS IN ROWS MARKED 1, 2, AND 4, SO, 1 + 2 + 4 = 7.

WRITE THE NUMBERS FOUND AT THE TOP OF EACH ROW THAT THESE NUMBERS APPEAR IN. THEN TOTAL THEM. DO YOU SEE ANYTHING INTERESTING?

15, 1 + 2 + 4 + 8 = 15   **11,**

14,   **8,**

13,   **6,**

12,   **5,**

58

ADDITION TEST

AFTER TAKING THE ADDITION FACT TEST THE STUDENT CAN MAKE A SET OF FLASH CARDS BY CUTTING THE TEST APART ALONG THE DARK LINES.

11 + 1	14 + 1	7 + 4	14 + 4
13 + 5	15 + 0	4 + 1	10 + 10
13 + 9	11 + 0	11 + 9	10 + 5
16 + 3	19 + 4	5 + 6	17 + 4
19 + 5	7 + 6	8 + 7	14 + 7
11 + 3	8 + 8	10 + 1	11 + 11
3 + 0	8 + 5	14 + 5	8 + 4
11 + 6	6 + 9	6 + 4	4 + 0
18 + 5	9 + 8	11 + 5	7 + 7

ADDITION TEST CONTD.

20+3	9+3	10+2	19+1
11+4	20+4	9+5	20+5
10+6	11+7	17+7	20+1
14+11	14+9	12+9	11+8
6+0	15+3	23+0	17+6
15+4	13+3	22+3	10+4
5+7	8+6	10+7	16+1
2+4	16+2	4+3	3+3
16+4	5+5	16+7	15+5

Excellent!

_____

YOU HAVE COMPLETED
YOUR ADDITION FACTS
TEST IN LESS THAN
60 SECONDS.

_____

Good Work, _____

 YOUR PAPER WAS 100% CORRECT.

 YOUR PAPER WAS VERY NEAT.

 YOUR WORK IS MUCH BETTER.

 YOU ARE LEARNING THE BASIC FACTS.

_____

SUPER!

_____

YOU HAVE PASSED THE
ADDITION FACTS TEST
IN LESS THAN 5 MINUTES.

_____

FANTASTIC

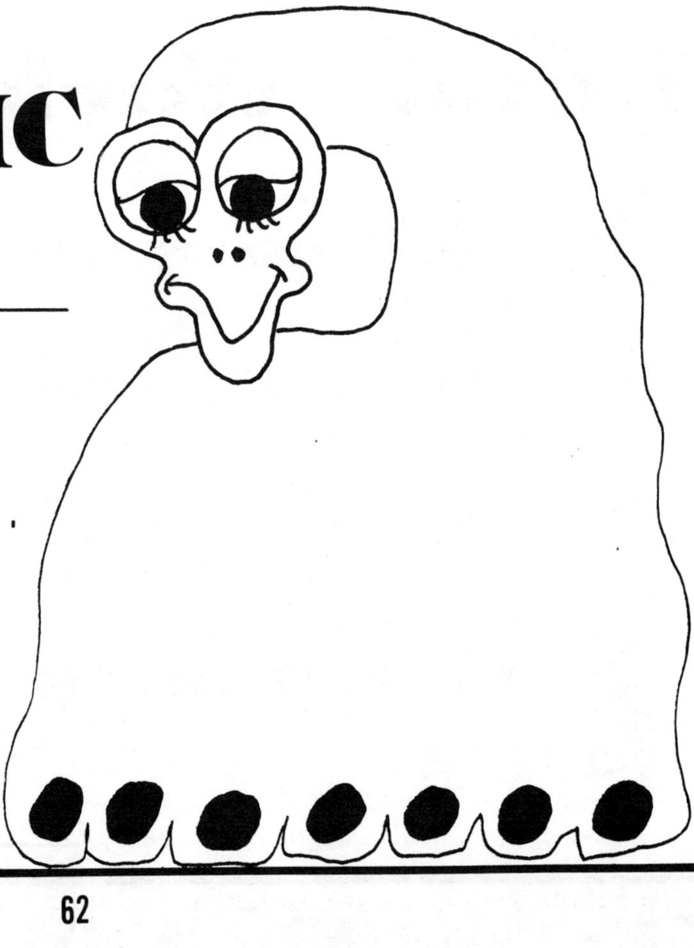

_____

YOU HAVE PASSED THE
ADDITION FACTS TEST
IN LESS THAN 3 MINUTES.

_____